29th Annual Saudi-Japan Symposium on Technology in Petroleum Refining and Petrochemicals 2019

Dhahran, Saudi Arabia
3-4 November 2019

ISBN: 978-1-7138-0312-6

Printed from e-media with permission by:

Curran Associates, Inc.
57 Morehouse Lane
Red Hook, NY 12571

King Fahd University of Petroleum & Minerals
Dhahran 31261
Kingdom of Saudi Arabia

Phone: +966 (13) 860 0000

info@kfupm.edu.sa

Additional copies of this publication are available from:

Curran Associates, Inc.
57 Morehouse Lane
Red Hook, NY 12571 USA
Phone: 845-758-0400
Fax: 845-758-2633
Email: curran@proceedings.com
Web: www.proceedings.com

الندوة السنوية السعودية اليابانية التاسعة والعشرون

29th Annual Saudi-Japan Symposium

التقنية في تكرير البترول والبتروكيماويات

Technology in Petroleum Refining & Petrochemicals

PROCEEDINGS: 29

7-6 ربيع الأول 1441هـ

November 3-4, 2019

KFUPM Research Institute

King Fahd University of Petroleum & Minerals

Dhahran, Saudi Arabia

Preface

This is the Proceedings of the 29th annual symposium "*Technology in Petroleum Refining & Petrochemicals*" held on November 3-4, 2019 at King Fahd University of Petroleum & Minerals (KFUPM), Dhahran, Saudi Arabia. This event, jointly organized by KFUPM, Saudi Aramco, the Japan Petroleum Institute (JPI), and Japan Cooperation Center, Petroleum (JCCP), brings together researchers from industry and academia to discuss recent downstream technologies related to refining and petrochemicals.

At this year's symposium, there are 23 presentations featuring three keynote papers addressing technical trends in Japanese refining, zeolites and catalysts in modern refinery. The additional 20 lectures focus on FCC olefins, HDS, crude distillation, dealkylation of aromatics, cetane improvement, production of C4 chemicals, polyethylene crystallization, methane reforming, hexane cracking, alkylation of USY zeolites, naphtha reforming, machine learning for yield prediction, cracking of pentene and crude oil to chemicals. Speakers are from Saudi Aramco, KFUPM, JXTG Nippon, JGC, Honeywell UOP, Japan AIST, Axens and Universities of Tottori, Tohoku and Utrecht.

We extend our thanks to KFUPM, Saudi Aramco, JPI, JCCP, distinguished speakers and attendees for their active participation and look forward to your participation in our next symposium.

Dr. Hassan S. Alasiri
Director, Center for Refining & Petrochemicals
http://ri.kfupm.edu.sa/cr
KFUPM-Research Institute
Dhahran 31261, Saudi Arabia

Contents

"Technical trends and research topics in the Japanese refining industry"

Demand for fuel oil is expected to fall by half in Japan, and will also decline in Asia. Therefore, production conversion from fuel products to chemical materials and olefins must be considered. The Japanese refining industry must approach two general issues. Firstly, the research and development business using refining technology must be promoted in the following areas. First, methods for manufacturing petrochemical materials must be improved. Direct production of the target materials and improved yield of the products such as olefin and butadiene are required with specific and efficient methods such as the separation membrane without depending on naphtha cracking. Petroleomics technology may be adapted to the structural analysis of heavy type oil and production technology. Second, FCV and hydrogen production will be developed to support transportation methods using liquid hydrogen, organic hydrides and ammonia. Such current development and direct electrolytic reduction are described as innovative technology. Secondly, renewable energy sources such as wind power, solar power and geothermal power will be promoted. On the other hand, research and development of CCS and CO2 utilization will continue to reduce CO2 emissions.

An Economic Perspective on Light Naphtha Upgrading

Yaming Jin and Omer R. Koseoglu (Refining Development/Saudi Aramco R&DC)

Light naphtha (C5-C6) streams, such as paraffinic light straight run naphtha (LSRN) or light cracked naphtha (LCN) from FCC units, have always been low-value hydrocarbon feedstocks. Steam cracking historically accounts for most of the world consumption of light naphtha to produce ethylene and propylene. With the booming shale gas production in America, more and more steam crackers are retrofitted to include additional capacity for ethane and propane feedstocks. More light naphtha feedstocks are looking for alternative upgrading routes.

In Saudi Arabia, the primary use for light naphtha is either direct blending into motor gasoline or isomerization for octane enhancement prior to blending. This approach is limited by the need for meeting gasoline RVP specification. A large amount of light naphtha is available in Saudi Aramco refineries and NGL fractionation plants.

To identify the most suitable route to upgrade light naphtha streams to value-added products, several alternative light naphtha conversion process schemes such as aromatization to produce benzene, toluene, and xylene (BTX), dehydrogenation followed olefins cracking process to produce propylene, dehydrogenation followed by dimerization to produce high quality gasoline blending components, etc., have been investigated. In this paper, we will present results of economic analysis of the potential process options by comparing gross margins of the processes of interest against those of steam cracking and hydroisomerization.

Recent Advances in Zeolite Catalysis for Production of Petrochemicals

Suheil F. Abdo, Ph. D.

Honeywell-UOP

Abstract

A major shift in the upgrading of petroleum resources from fuels to petrochemical production is currently underway. Led by Aramco and other major producers, this shift is taking place due to the anticipated decrease in demand for fossil fuels resulting from environmentally driven enhancements in fuel economy, electrification and the growing deployment of renewable energy sources such as wind, solar and bio renewables. Key initiatives including Crude-to-Chemicals and Refining-Petrochemical Integration or Refinery of the Future have generated great interest across the globe as evidenced by many announced projects. Interestingly, both crude-to-chemicals approaches. Both of these approaches include processing steps that would benefit critically from precise molecular management with the goals of converting feed and intermediate molecules to the desired light olefins and BTX aromatics. Consequently, to realize the long-range scenario depicted in the figure below, technology developers must rely on processes that allow precise control of molecular transformations to deliver the highest value products. Thermal and catalytic processes are typically proposed to deliver and the latter invariably include zeolite based technologies.

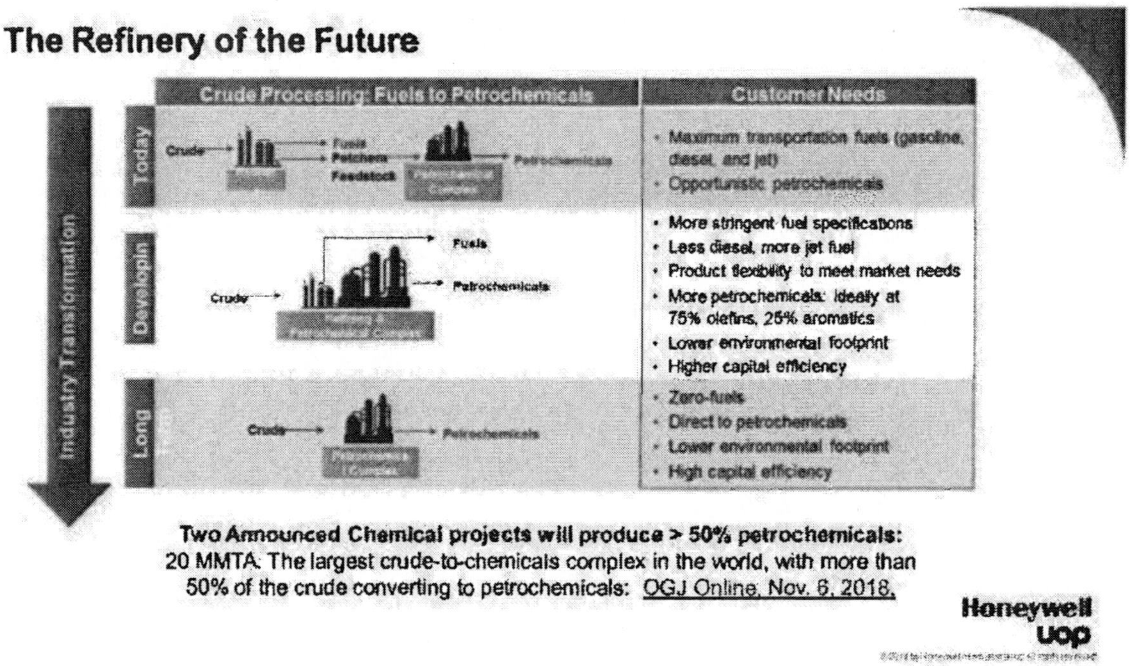

The flow scheme below illustrates how an integrated fuels and petrochemical complex might be configured to deliver the optimum value from crude oil processing. Zeolite based technologies such as FCC, Hydrocracking and Alkylation are key to delivery of the desired mix of products with the highest value. Fundamentally, the most successful offerings are based on careful control of key catalyst properties including appropriate zeolite geometry and acid strength coupled with the appropriate process conditions to promote desired cracking and hydrogen transfer reactions while suppressing formation of coke and other secondary products that lead to rapid catalyst deactivation.

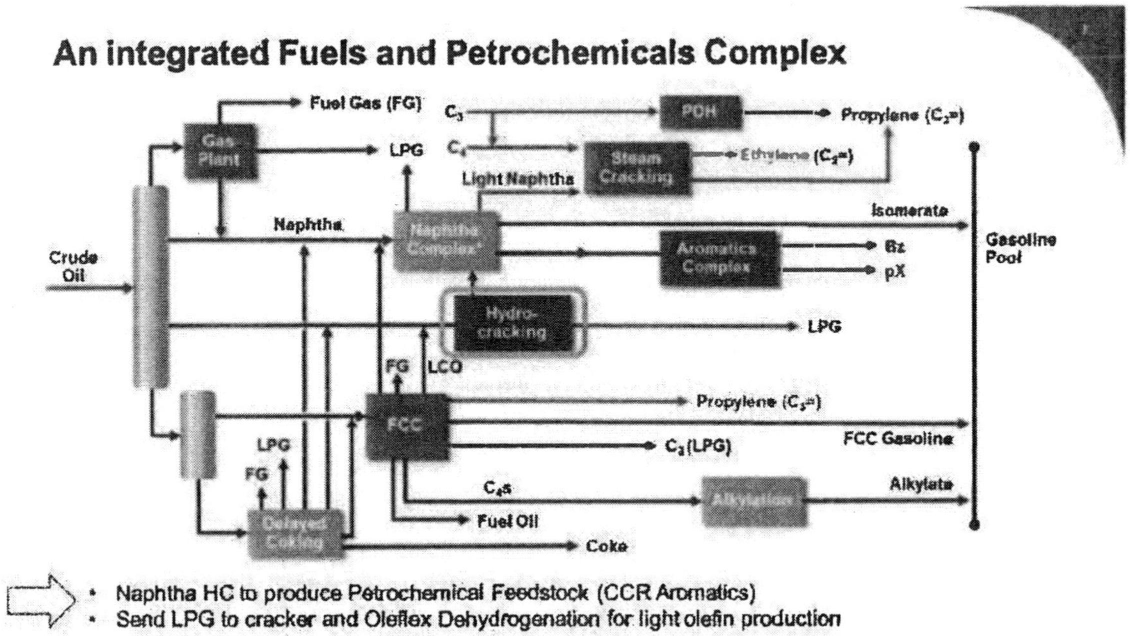

This is accomplished by implementing catalyst design elements which consider relative diffusivities of various feed components (molecules) and their varying affinities for catalytic sites. Coupling of the optimum support porosity and geometric pore structure with acidity and other surface properties will be critical to development of the appropriate catalyst system.

This presentation will offer an overview of the major trends in refining, petrochemical and energy industries and review the rich history and current trends in zeolite development thereby illustrating their critical role in meeting the evolving needs of our industry as well as society in general. It will also show how combining synthesis and modification of new zeolite frameworks with state of the art characterization tools have yielded a significant fundamental understanding of the variables impacting catalytic behavior and have enabled successful application in a range of petrochemical, refining and environmental technologies.

Catalyst Technology for Maximize Light Olefine Yield in FCC-U

Takaki Mizuno
FCC Catalyst Research Group, Petroleum Refining Catalysts Research Center,
JGC Catalysts and Chemicals Ltd.
13-2, Kitaminato-machi, Wakamatsu-ku, Kitakyushu-shi, 808-0027 Japan
E-mail: mizuno.takaki@jgccc.com

Abstract

In the present work, we have investigated how the component and carbon number distribution of gasoline products on FCC catalyst would change with ZSM-5 additive, the effects of the host FCC catalysts reactivity and metals tolerance of ZSM-5 additive. Addition of ZSM-5 additive has converted gasoline fraction to light olefins, but it has not converted LCO and HCO. ZSM-5 additive has decreased C_5-C_8 olefins and iso-paraffins in FCC gasoline components and would increase C_2-C_4 olefins and iso-C_4 paraffin, while aromatics have been remained almost unchanged. In the case of changed RE_2O_3 content in the host FCC catalyst, the propylene yield has been increased against conversion and gasoline yield on the low RE_2O_3 content. As using the host FCC catalyst with high RE_2O_3 content, high carbon number olefins would decrease, while aromatics would increase due to hydrogen transfer reaction. So that propylene yield would decrease. The effect of metals on the ZSM-5 additive performance has been investigated. In the case of impregnated Ni on ZSM-5, the propylene yield has been hardly affected, while it has been decreased in the impregnated V case. Therefore, we have developed a new ZSM-5 additive, "OCTUP-R" that applied the V tolerance matrix technology. The performance of OCTUP-R has been kept higher even with high metals contents as compared with that of current ZSM-5 additive, OCTUP-α. The number of acid sites, crystallinity of ZSM-5 zeolite and zeolite surface area of OCTUP-R with high metals content has been kept higher level as compared with those of OCTUP-α due to applied V tolerance matrix technology. It is thought that OCTUP-R is the effective for use in RFCC-U.

1. Introduction

Light olefins such as propylene and butene are basic building blocks for the manufacture of petrochemical products. In particularly, the demand for propylene has increased over the last decade and is expected to increase worldwide (shown in Fig. 1). There are some processes for

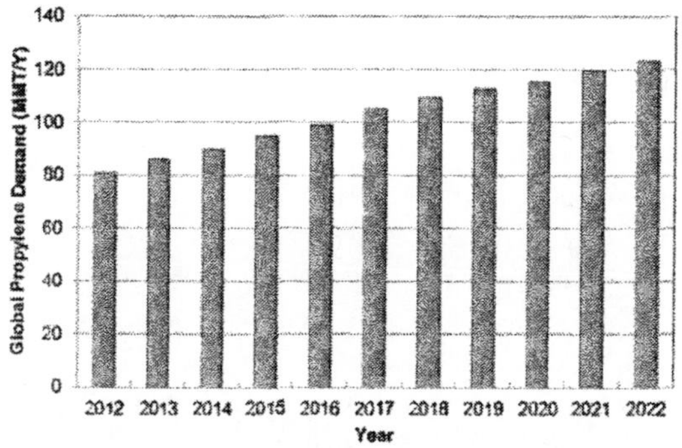

Fig. 1 Global demand of Propylene

propylene production, for examples steam cracker, propane dehydrogenation, methanol to olefins. One of the major sources of propylene is from Fluid Catalytic Cracking (FCC) process. Conventional propylene yield in FCC unit is 3-7 wt%, however, high propylene yield FCC process, such as High Severity FCC (HS-FCC), Deep Catalytic Cracking (DCC), etc., has been developed and the propylene yield would be reached 15-20 wt%.

In terms of the catalyst in FCC-U, ZSM-5 additive have been blended into host FCC catalyst for increasing light olefins yield [1, 2]. JGC Catalysts and Chemicals Ltd. (JGC C&C) have developed and commercialized various advanced ZSM-5 additive as "OCTUP series" to meet refiner's requirements. Table 1 shows JGC C&C's ZSM-5 additive. "OCTUP-R", the latest grade, has been developed for resid FCC, because the number of RFCC units for propylene mode operation has been increasing in the Southeast Asia. It is improved in metal (V) tolerance. It is indicated that the performance test and the characterization results in this paper.

Table 1 JGC C&C's ZSM-5 additives "OCTUP series"

Name	Feature	ZSM-5 content (%)
OCTUP-11S	Standard Type, C3= & RON up	25
OCTUP-α	High Activity Type, C3= & RON MAX	40
OCTUP-C4	C4= Selective, Less C2=, Higher RON BLL	40
OCTUP-R	High Activity for Resid FCC, C3= &RON MAX	40

In order to increase the propylene yield, even if the content of ZSM-5 additive is increased, propylene yield reach the limit, so that it is important that development including the host FCC catalyst. The design of host catalyst affects not only products yield but also gasoline component. The analysis of gasoline components is very important for understanding the function of ZSM-5 additive and the developing the host FCC catalyst. It is analyzed that the gasoline components of ACE-MAT evaluations in cases of base catalyst (CVZ-1) with/without ZSM-5 additive (OCTUP-α) and different designed catalyst (CVZ-2A, 2B and 2C) with OCTUP-α. It is discussed that the function of ZSM-5 additive and the effect of FCC catalyst design on the light olefins yield in this work.

2. Experimental

2.1. Feed properties and FCC catalyst, ZSM-5 additive

The feed A, which is mainly heavy vacuum gas-oil (HVGO), was used in section 3 and 4. The feed B, which is desulfurized vacuum gas-oil (DSVGO), was used in section 5. The properties of feed are given in Table 2. FCC catalysts

Table 2 Feed properties

Feed		A	B
Density	(g/ml)	0.8910	0.9040
K factor	(-)	12.15	12.12
Sulfur	(wt%)	0.42	0.21
Nitrogen	(wt ppm)	725	800
Con. Carbon	(wt%)	0.18	0.34
Distillation 10%	(°C)	368	385
Distillation 50%	(°C)	433	457
Distillation 90%	(°C)	507	546
Molecular Weigh	(-)	409	445

named CVZ-1 and CVZ-2A, 2B and 2C, ZSM-5 additive OCTUP-α and OCTUP-R were tested in this work. Physical and chemical properties of catalysts and ZSM-5 additive are given in Table 3.

Table 3 Catalyst properties

Catalyst		CVZ-1	CZV-2A	CVZ-2B	CVZ-2C	OCTUP-α	OCTUP-R
RE_2O_3		middle	low	middle	high	-	-
Al_2O_3	(wt%)	46	38	38	38	25	30
Total surface area	(m^2/g)	205*	206*	208*	206*	115	124
Zeolite surface area	(m^2/g)	123*	130*	131*	134*	-	-
Matrix surface area	(m^2/g)	82*	76*	77*	72*	-	-
Zeolite unit cell size	(Å)	24.34*	24.30*	24.32*	24.33*	-	-

*After Deactivation with 5% OCTUP-α, Deactivation: metal free, calcined with 100% steam at 780°C for 13hr

2.2. Catalyst pre-treatment

The FCC catalysts were deactivated by calcination with 100 % steam at 780 °C for 13 hr. In the case of OCTUP-R, it was impregnated to Ni ppm/V ppm: 1,000/2,000, 2,000/4,000, 3,000/6,000 and 5,000/10,000 and deactivated by calcination with 100 % steam at 810 °C for 13 hr. After then, it was blended to equilibrium catalyst.

2.3. Reaction test and analysis of products

The catalyst reaction test was carried out with a fixed-fluidized micro activity (ACE-MAT) unit and reacted at 547 °C and Dry gases (H_2, C_1-C_2), LPG (C_3-C_4), gasoline (C_5 < bp < 220 °C), light cycle oil (LCO) (220°C < bp < 370°C) and heavy cycle oil (HCO) (bp > 370°C) yield were calculated from GC analysis in section 3 and 4. The reaction test in section 5 was carried out at 510 °C and the cut points were as follows; gasoline: C_5 < bp < 216 °C, LCO: 216°C < bp < 343°C, HCO: bp > 343°C. Coke yield was calculated by measured CO_2 production during the complete coke oxidation. The conversion was defined as followed; 100 wt% - (LCO (wt%) + HCO (wt%)).

3. Influence of added ZSM-5 additive

The results of catalyst reaction tests were shown in table 4. In the case of OCTUP-α addition, it was added 5 wt%. As adding OCTUP-α, the LPG yield, in particularly propylene increased instead of the gasoline yield decreasing. On the other hand, the yields of LCO and HCO were almost the same. These results mean that the function of OCTUP-α is convert gasoline to light olefins such as C_3=, C_4= and not to convert LCO and HCO hardly.

The relationships of products (gasoline + LPG +C_2) yield and distributions of carbon number were shown in Fig. 2. As adding OCTUP-α, C_5-C_8 compounds in total yield decreased, while C_2-C_4 compounds increased (Fig. 2a). The low carbon number ($\leq C_4$) compounds of olefins and iso-paraffins increased, while the high carbon number ($\geq C_5$) compounds of them decreased with OCTUP-α addition (Fig. 2b and c). On the other hand, the carbon number distributions in aromatics were almost the same in both cases

Table 4 Results of ACE-MAT evaluation

Catalyst		CVZ-1		CVZ-1 with OCTUP-α	
C/O	(-)	3.75	5.00	3.75	5.00
Conversion	(wt%)	84.02	85.68	83.95	85.53
H_2	(wt%)	0.09	0.08	0.08	0.08
C_1	(wt%)	1.0	1.0	0.9	0.9
C_2	(wt%)	0.6	0.6	0.6	0.6
C_2=	(wt%)	0.8	0.9	1.7	1.9
C_3	(wt%)	1.6	1.7	2.4	2.5
C_3=	(wt%)	7.5	7.9	13.0	13.2
iC_4	(wt%)	5.8	6.3	7.1	7.7
n-C_4	(wt%)	1.4	1.4	1.3	1.4
C_4=	(wt%)	9.8	9.9	11.8	11.6
Gasoline	(wt%)	51.8	51.5	41.4	41.3
LCO	(wt%)	11.9	10.8	11.8	10.9
HCO	(wt%)	4.1	3.5	4.2	3.6
Coke	(wt%)	3.7	4.4	3.7	4.3

Rx. Temp.: 547 °C, C/O: 3.75 & 5.00, Feedstock: A
Cut point: gasoline: C_5~220°C, LCO: 220°C~370°C, HCO: 370°C<

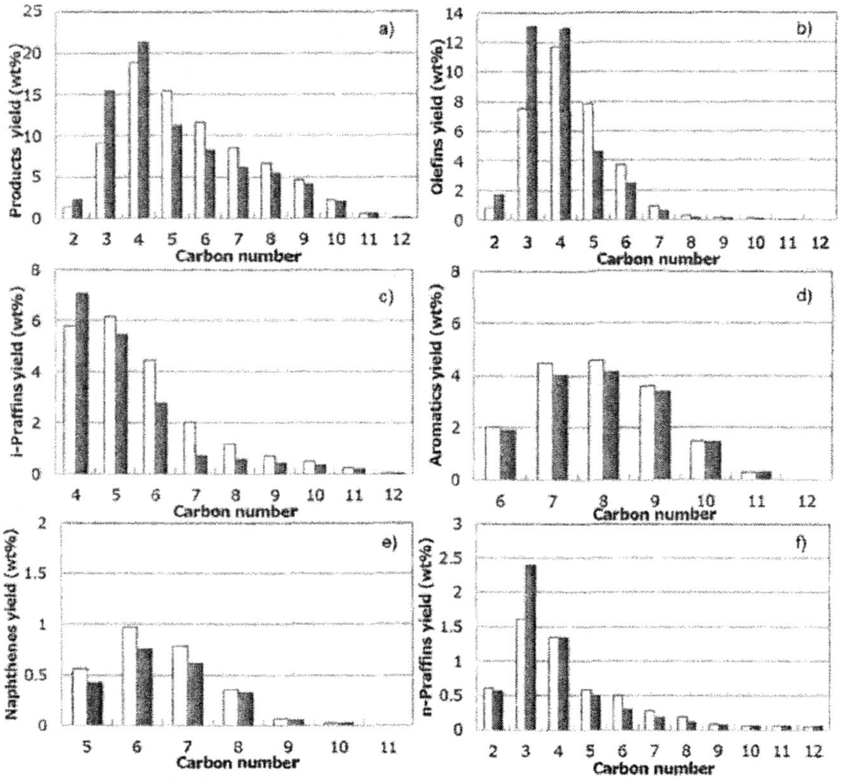

Fig. 2 Product yield versus carbon number at C/O: 3.75, a) total (C_2 +LPG + gasoline) yield, b) olefins yield, c) iso-paraffin yield, d) aromatics yield, e) naphtens yield and f) n-paraffins, □: CVZ-1, ■: CVZ-1 with OCTUP-α (5wt%)

(Fig. 2d). The naphthens yield and the n-paraffins yield were less than others (Fig. 2e and f), so that it is thought that added OCTUP-α influence the carbon number distributions in them hardly without propane yield. The fact that the propane yield was increased is thought to be formed by the cracking iso-paraffins on OCTUP-α. Therefore, these results mean that the high carbon number (≥ C_5) of olefins and iso-paraffins are mainly converted to light oleffins (≤ C_4) and iso-C_4 on OCTUP-α. On the other hand, it could not crack aromatics.

4. Influence of host FCC catalyst performance

In order to increase light olefins, such as propylene, butene from the improvement of the catalyst, it is important to increase the number of C_5-C_9 olefins and iso-paraffins on the host FCC catalyst. In generally, to increase them, the hydrogen transfer reactions as shown by the following: 4Olefins → 3Paraffins + Aromatic, Olefin + Naphthene → Paraffin + Aromatic, has to be suppressed on FCC catalyst [3-5]. For this purpose, it is adopted that the low rare earth oxide (RE) to zeolite ratio FCC catalyst. In this section the different RE/zeolite FCC catalysts (CVZ-2A, CVZ-2B and CVZ-2C) with OCTUP-α are tested and it is compared that the relations light olefins yield and each gasoline component yield.

The results of reaction tests in the different RE_2O_3 content in FCC catalysts were shown in table 5. The conversion and gasoline yield increased with increasing RE_2O_3 content. On the other hand, the propylene and butene yield decreased with increasing RE_2O_3 content. These catalysts have almost the same zeolite and matrix surface area. Therefore, these results mean that the difference in the products yield is depending on RE_2O_3 content. It is known that high RE_2O_3 to zeolite ratio, such as RE-USY having high UCS, would induce high acid site density after the hydrothermal deactivation [3, 5]. The catalyst having high acid site density would cause hydrogen transfer reaction as the secondary reaction. In this case, it is suggested that there is a difference in the acid

Table 5 Results of ACE-MAT evaluation

Catalyst		CVZ-2A		CVZ-2B		CVZ-2C	
RE_2O_3		low		middle		high	
ZSM-5 additive	(wt%)	5		5		5	
C/O	(-)	3.75	5.00	3.75	5.00	3.75	5.00
Conversion	(wt%)	82.03	83.99	83.50	85.38	84.12	85.72
H_2	(wt%)	0.07	0.06	0.07	0.06	0.06	0.06
C_1	(wt%)	0.8	0.9	0.8	0.9	0.9	0.9
C_2	(wt%)	0.5	0.5	0.6	0.5	0.6	0.6
C_2=	(wt%)	1.7	1.9	1.6	1.7	1.6	1.7
C_3	(wt%)	2.2	2.4	2.4	2.5	2.6	2.8
C_3=	(wt%)	13.7	14.1	12.7	13.0	12.1	12.3
iC_4	(wt%)	6.2	6.9	7.0	7.6	7.7	8.3
n-C_4	(wt%)	1.2	1.3	1.3	1.4	1.5	1.5
C_4=	(wt%)	12.7	12.7	11.8	11.6	10.9	10.7
Gasoline	(wt%)	39.8	39.5	42.0	42.0	42.5	42.3
LCO	(wt%)	12.5	11.4	11.8	10.7	11.3	10.4
HCO	(wt%)	5.5	4.6	4.7	3.9	4.6	3.9
Coke	(wt%)	3.1	3.8	3.3	4.0	3.8	4.5

Rx. Temp.: 547 °C, C/O: 3.75 & 5.00, Feedstock: A

Cut point: gasoline: C_5~220°C, LCO: 220°C~370°C, HCO: 370°C<

site density at the range of these RE_2O_3 amounts.

The carbon number distributions of products were shown in Fig. 3. The carbon numbers of C_5-C_9 compounds in the products were decrease with the lower RE_2O_3 content (Fig. 3a). It means that the catalyst, having high RE_2O_3 content, indicated high gasoline yield due to increasing high carbon number compounds. The propylene and butene yields were tended to increase with the decreasing RE_2O_3 content (Fig. 3b). The yields of higher carbon number ($\geq$ C5) olefins on CVZ-2A and CVZ-2B were almost the same, while those on CVZ-2C were decreased. These results suggested that the difference in the propylene and butene yields was not depending on ZSM-5 additive performance but depending on the difference in higher carbon number olefins yield on the host FCC catalysts. Namely, at the higher RE_2O_3 content, the reason of decrease in propylene and butene yields is decreasing their sources, which are higher carbon number olefins. These results are similar to the literature using RE-USY [6, 7]. The aromatics (Fig. 3d), iso-paraffins (Fig. 3c) and n-paraffins (Fig. 3f) increased with the

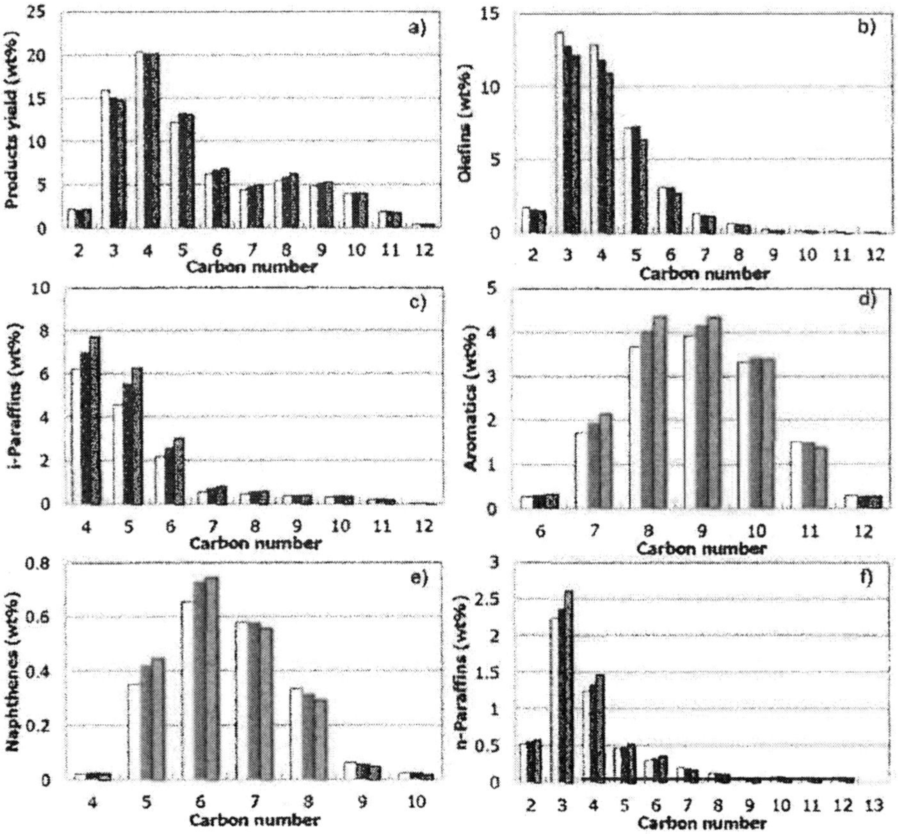

Fig. 3 Product yield versus carbon number at C/O: 3.75, a) total (C_2 +LPG + gasoline) yield, b) olefins yield, c) iso-paraffin yield, d) aromatics yield, e) naphtens yield and f) n-paraffins, □ : CVZ-2A, ■ : CVZ-2B, ▨ : CVZ-2C

increasing RE_2O_3 content. As shown in Fig. 2d, aromatics could not change with addition of ZSM-5 additive, so the difference in the aromatics yields is depending on the difference in the activity of hydrogen transfer reactions on the host FCC catalysts. The facts that iso-paraffins and n-paraffins were increasing with increasing RE_2O_3 contents support it. In the case of naphthenes (Fig. 3e), although the yield is a quite low as compared with others, C_5-C_6 compounds tended to be higher yields with the higher RE_2O_3 content.

Focusing on the propylene yield, these results suggest that a lower RE_2O_3 content is desirable, but conversion and gasoline yield decrease. The yield of each product such as gasoline would affect the economics in the actual unit. In addition, the actual FCC units have constraints such as recovery of dry gas. Therefore, it is important to consider a combination of the host FCC catalyst and ZSM-5 additive to maximize the profit of each unit.

5. High metal tolerance ZSM-5 additive

It is known that metals, which are included in rasid feed, have the catalyst performance decrease in RFCC-U due to broken zeolite structure [8-10]. However, the effect of metals on the ZSM-5 additive performance is not well known, because the amount of added ZSM-5 additive is quite less than that of FCC catalyst. Therefore, the effect of metals on ZSM-5 additive performance is studied by Ni, V impregnated on ZSM-5 additive alone in this work.

It is shown in Fig. 4 that the relationship between propylene yield and additive amount when OCTUP-α is impregnated with Ni or V. Even when 4,000 ppm of Ni was loaded on OCTUP-α, the propylene yield was almost unchanged from that of metal-free. On the other hand, in the case of V, at the same additive amount propylene yield clearly decreased, and it is necessary to increase the additive amount in order to obtain the same propylene yield as to that of metal-free. These results suggest that V species on ZSM-5 additive caused to break the ZSM-5 zeolite structure, while Ni species could not cause to that. Therefore, it is important to improve the V tolerance in order to suppress the

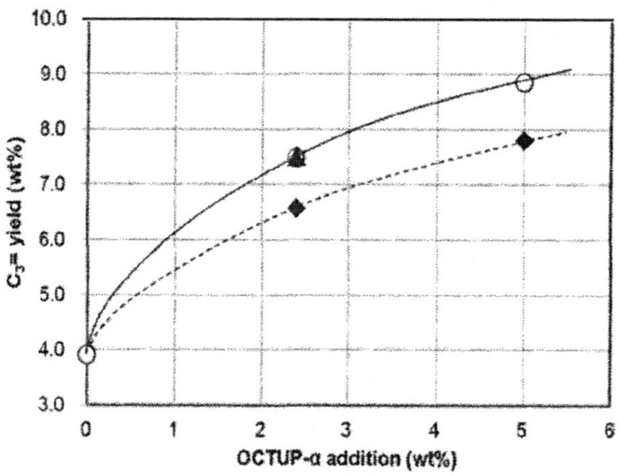

Fig. 4 The effect of Ni, V on OCTUP-α for propylene yield Reaction condition: C/O = 5.0, Rx. Temp. = 510 °C, Feedstock : B, Cut point: gasoline: C_5~220 °C, LCO: 220 °C~370 °C, HCO: 370 °C <, metal content on OCTUP-α, ○: metal free, ▲: Ni 4,000 ppm, ◆: V 4,000 ppm

performance degradation of the ZSM-5 additive. In order to improve the V tolerance of FCC catalyst, JGC C&C possesses CMT-40 as a matrix technology, and we thought that applying this technology to ZSM-5 additive would improve the V tolerance. Based on such catalyst design, a new ZSM-5 additive, "OCTUP-R" was developed.

The comparative results of propylene yield on OCTUP-R and OCTUP-α with various metals contents were shown in Fig. 5. In all range of Ni plus V, OCTUP-R indicated higher propylene yield as compared with OCTUP-α. The difference of them was widely at the range of 3,000-6,000 ppm of Ni plus V contents, while it decreased over 9,000 ppm of Ni plus V contents. This difference is thought to be due to the difference in the number of acid sites after deactivation.

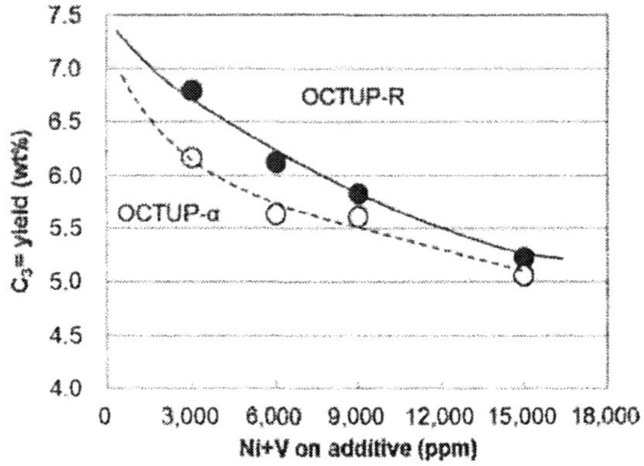

Fig. 5 The effects of metal (Ni + V) amounts on OCTUP-R for propylene yield as compared with OCTUP-α, Additive addition: 2.4wt% into E-cat., Reaction condition: C/O = 5.0, Rx. Temp. = 510 °C, Feedstock : B, Cut point: gasoline: C_5~220 °C, LCO: 220 °C~370 °C, HCO: 370 °C <,
●: OCTUP-R, ○: OCTUP-α

The numbers of acid sites were measured by Temperature programmed desorption (TPD) of adsorbed ammonia and pyridine adsorbed FT-IR. The results of fresh and deactivated catalysts with Ni/V = 2,000ppm/4,000ppm were shown in Table 6. Although the number of acid sites differs depending on the analytical method, the number of acid sites on OCTUP-R was about 1.4 times higher for the fresh catalyst than that on OCTUP-α. The difference spread about twice for the deactivated catalysts. These results mean that the number of acid sites on OCTUP-R was maintained higher than that on OCTUP-α. It suggests that the applied V tolerance matrix technology contributed.

Table 6 The number of acid sites on OCTUP-R and OCTUP-α

Catalyst		OCTUP-R	OCTUP-α
The number of acid sites on the fresh catalysts			
NH_3 adsorption	(μmlo/g)	229	171
Total pyridine adsorption	(μmlo/g)	41	28
Brønsted acid sites	(μmlo/g)	30	25
Lewis acid sites	(μmlo/g)	10	3
The number of acid sites on the deactivated catalysts with Ni/V=2,000ppm/4,000ppm			
NH_3 adsorption	(μmlo/g)	51	24
Total pyridine adsorption	(μmlo/g)	12	6
Brønsted acid sites	(μmlo/g)	5	3
Lewis acid sites	(μmlo/g)	7	3

From results of pyridine adsorption FT-IR, Brønsted acid sites and Lewis acid sites are separated. The numbers of Brønsted acid sites, which are on the ZSM-5 zeolite, in the fresh catalysts were not so different. However, after deactivation with metals, they were markedly decreased and the number of Brønsted acid sites on OCTUP-R was more than that on OCTUP-α. The difference in the number of Lewis acid sites is probably based on the difference in the added matrix components and they are considered not to contribute to the cracking reaction of gasoline component to light olefins. Therefore, it is suggested that the difference in the number of Brønsted acid sites may affect the performance difference of ZSM-5 additive.

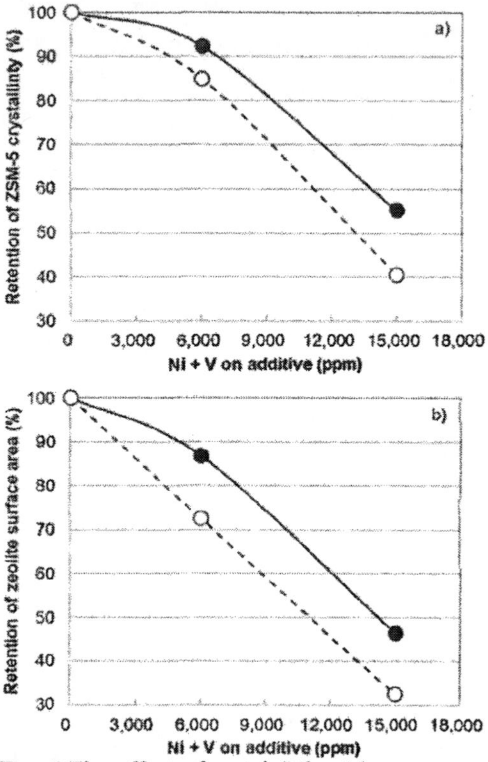

Fig. 6 The effect of metal (Ni + V) amount on ZSM-5 crystallinity and zeolite surface area for OCTUP-R and OCTUP-α, ●: OCTUP-R, ○: OCTUP-α

As other factors affected by the metals on ZSM-5 additive, the retention of ZSM-5 zeolite crystallinity and zeolite surface area were shown in Fig. 6. The crystallinity of OCTUP-R was kept higher at any metals content as compared with OCTUP-α. The zeolite surface area of OCTUP-R was also kept higher. These results show the same results as the number of acid sites. Therefore, it is concluded that OCTOUP-R has higher metal tolerance than OCTUP-α. OCTUP-R with high V tolerance is effective for use in RFCC-U, and it is possible to reduce the additive addition rate even at high metal on catalysts.

6. Conclusions

The functions of ZSM-5 additive was studied using OCTUP-α. The ZSM-5 additive converted gasoline to light olefins such as C_3=, C_4= but it did not convert LCO and HCO. In terms of gasoline components, the higher carbon number of olefins and iso-paraffins were converted to propylene, butene and iso-butane on OCTUP-α, while aromatics were not changed. It is important to increase the number of C_5-C_9 olefins and iso-paraffins on host FCC catalyst for maximize the light olefins yields.

The effects of RE_2O_3 contents in the FCC catalysts on the light olefins yields were investigated. The light olefins yields increased on lower RE_2O_3 content in catalysts, while the conversion and gasoline yield decreased. Increased RE_2O_3 content in the FCC catalyst promoted the hydrogen transfer reactions and reduced the yield of the high carbon number olefins, which are light olefins sources on ZSM-5 additive. However, the low RE_2O_3 content in the FCC catalyst would induce the decrease in the conversion and liquid recovery, so that it is important that to take into consideration not only the yield of light olefins but also the overall product yields and some constraints in actual units.

The performance of ZSM-5 additive decreased due to deactivation with impregnated V in this work. New catalyst OCTUP-R with improved V tolerance indicated higher propylene yield as compared with OCTUP-α in all range of Ni plus V amounts. Applied V tolerance technology suppressed the breaking the ZSM-5 zeolite structure by V species, and as a result the retention of the number of acid sites, crystallinity and zeolite surface area. It is thought that OCTUP-R is very effective for use in RFCC-U, and it is possible to reduce the additive addition rate even at high metal on catalysts.

7. References

[1] J. S. Buchanan, Catal. Today, **55**, 207 (2000)
[2] T. F. Degnan, G. K. Chitnis, P. H. Schipper, Microporous Mesoporous Mater. **35-36**, 245 (2000)
[3] Conghua Liu, Xionghou Gao, Zhongdong Zhang, et al., Appl. Catal. A, **264**, 225 (2004)
[4] S. Al-Khattaf, Appl. Catal. A, **231**, 293 (2002)
[5] D. Wallenstein, K. Schäfer, R. H. Harding, Appl. Catal. A, **502**, 27 (2015)
[6] Xinjin Zhao, Robert H. Harding, Ind. Eng. Chem. Res., **38**, 3854 (1999)
[7] D. Wallenstein, R. H. Harding, Appl. Catal. A, **214**, 11 (2001)
[8] Mahmoud M. Khader, J. Mol. Catal. A, **104**, 87 (1995)
[9] E. Tangstad, A. Andersen, E. M. Myhrvold, T. Myrstad, Appl. Catal. A, **346**, 194 (2008)
[10] F. V. Pinto, A. S. Escobar, B. G. de Oliveira, et al., Appl. Catal. A, **388**, 15 (2010)

Recent advances and overviews of Ni(Co)Mo supported mesoporous silica catalysts for HDS activity of liquid fuel.

Saheed A. Ganiyu, Khalid Alhooshani and Abdulkadir Tanimu

Chemistry Department, King Fahd University of Petroleum & Minerals, Dhahran 31261, Saudi Arabia.

Abstract

The advent of industrialization resulted from fast rising of human population and technological needs, has drastically increased the environmental pollution worldwide. For example, the uprising level of sulfur-containing compounds in the atmosphere due to emission from automobiles and hydrocarbon processing is of great concern that necessitate urgent social and scientific attentions. The current sulfur content in the transportation fuels should be as low as 10 ppm as promulgated by environmental policy makers, which is a stringent regulation on the petroleum refiners. The recent catalysts development for HDS activity employs Ni(Co)Mo supported mesoporous silica modified with different heteroatoms (Ti, Zr and Ce) and chelating agent. Our recent studies established that HDS catalytic performance of sulfided-Ni(Co)Mo is directly correlated with modification of the amorphous silica by chelating agent and heteroatoms. The catalytic properties with respect to specific surface area and pore-size distribution, ease of reduction, degree of sulfidation of metal oxides and dispersion of the active species are greatly improved by these strategies. The series of heteroatoms modified catalysts developed have great potential for ultradeep desulfurization.

HDS Performance Ti-modified SBA-15-NiMo Catalyst Single-pot Approach.

This strategy employs facile and low-cost approach to incorporate nickel and molybdenum into mesoporous silica framework by direct synthesis (single-pot) as shown in the scheme below. This involves synthesis of mesoporous silica modified with titanium and active species (Ni and Mo) in one pot) with or without complexing agent using Pluronic P123 as an organic template.

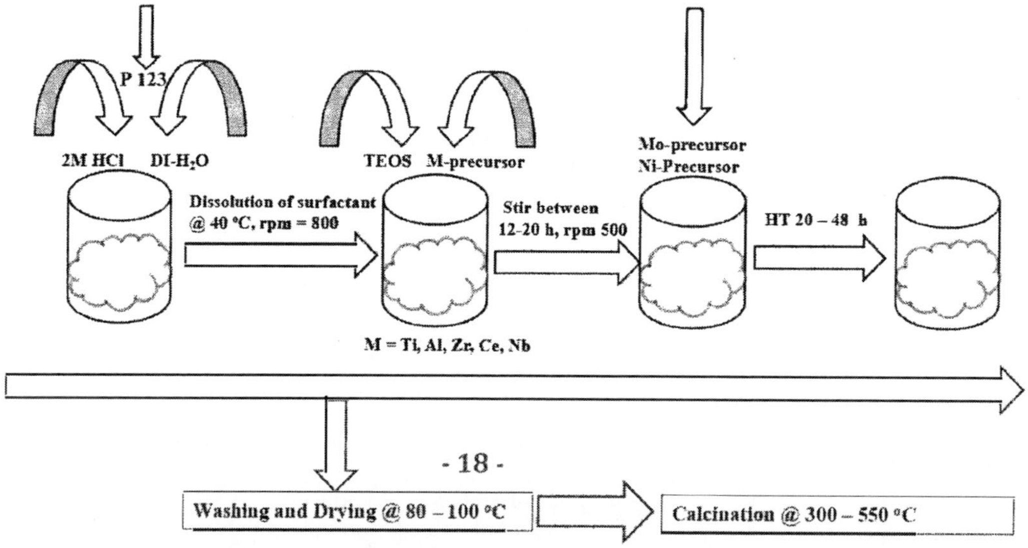

- 18 -

Fig. 1. Schematic Representation of Single-Pot (SP) Synthesis of NiMo Supported on Metal-Modified SBA-15 Catalysts [1].

The HDS performance of the as-synthesized catalysts were conducted in batch reactor at 350 °C and 5 MPa for dibenzothiophene (DBT). The catalytic properties of single-pot catalysts at low calcination temperature are preserved and resulted in better dispersion of molybdenum, higher acidity and accessibility of the pores. Thence, overall HDS is significantly improves and higher than conventional impregnation method.

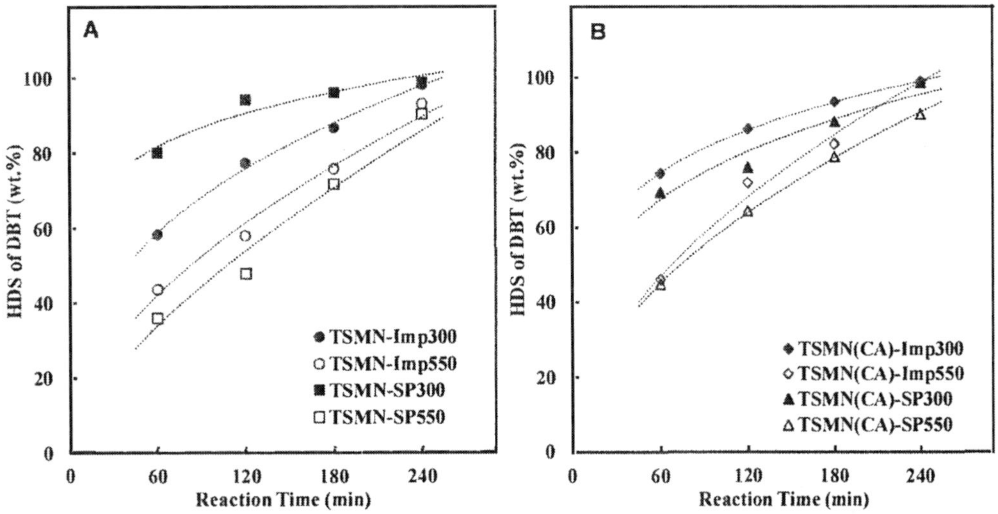

Fig. 2: HDS activity of catalysts prepared by different approaches [1].

The reaction pathway via direct desulfurization (hydrogenolysis) and hydrogenation were observed in the HDS reaction of DBT (Fig. 3). The single-pot catalyst favors direct cleavage of C–S to form biphenyl with low consumption of hydrogen than hydrogenation steps with multiple intermediates.

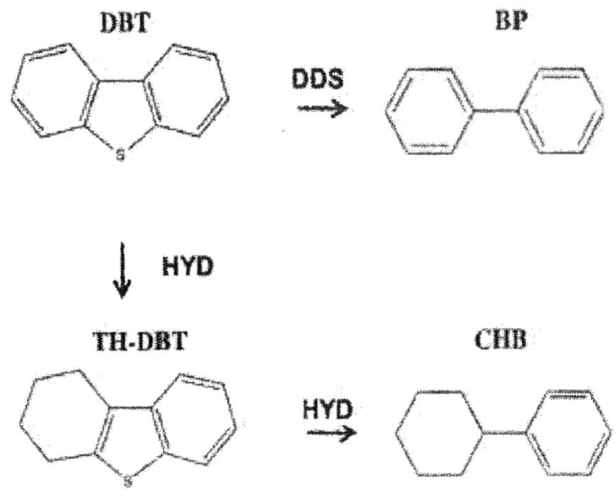

Fig. 3: Reaction pathway for HDS of DBT [1].

HDS Activity of Zr-modified SBA-15 NiMo Catalyst by One-step Calcination and Reduction

The insight into one-step calcination and reduction strategy was adopted to prepare better HDS catalysts for simultaneous HDS of DBT and DMDBT in batch reactor. Our studies revealed that multi-step heat treatment resulted in loss of catalytic properties and performance [2]. The characteristics investigation of one-step heat-treated catalysts confirmed that specific surface area, surface acidity and MoS_2 dispersion is greatly enhanced. The catalytic desulfurization performance for both DBT and DMDBT was better than catalysts prepared by conventional sequential heat treatment.

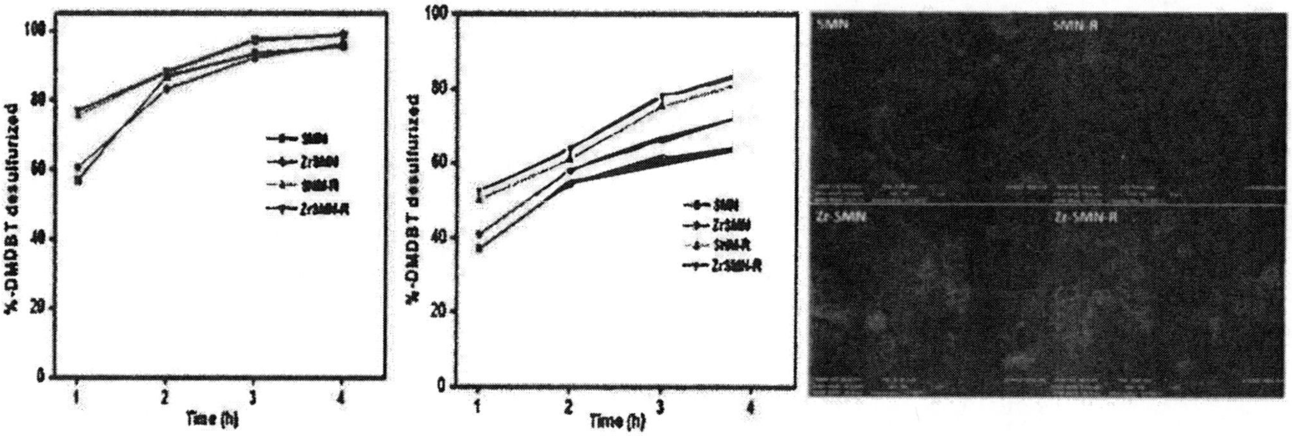

Fig. 4: Catalytic performance and SEM analysis of (Zr)-modified SBA-15-Mo/Ni catalysts with different heat treatment [2].

Conclusion

Modification of amorphous mesoporous silica having low support interaction with active species (Ni(Co)Mo) by heteroatoms (Ti, Zr or Ce) is a viable strategy towards highly dispersed and high degree of HDS catalysts to achieve ultradeep desulfurization. The introduction of single-pot synthesis strategy and addition of chelating agent improves the catalytic performance significantly.

References

[1] Ganiyu, Saheed A., Khalid Alhooshani, and Syed Ahmed Ali. "Single-pot synthesis of Ti-SBA-15-NiMo hydrodesulfurization catalysts: Role of calcination temperature on dispersion and activity." *Applied Catalysis B: Environmental* 203 (2017): 428-441.

[2] Ganiyu, Saheed A., and Khalid Alhooshani. "Catalytic Performance of NiMoS Supported on (Zr) SBA-15 for Hydrodesulfurization of Diesel: Insight into a One-Step Calcination and Reduction Strategy during Sulfidation." *Energy & Fuels* 33, no. 4 (2019): 3047-3056.

Effects of Titania Coatings on Hydrodesulfurization Catalysts: Insights from First-Principles Calculations

Hiroki Iriguchi[†1)], Hiroki Uratani[†2)], Takao Kudo[†1)], Hiroyuki Seki[†1)], and Koichi Yamashita[†2)]

[†1)]Central Technical Research Laboratory, JXTG Nippon Oil & Energy Co., 8 Chidoricho, Naka-ku, Yokohama, 231-0815, JAPAN

[†2)]Dept. of Chemical System Engineering, School of Engineering, The University of Tokyo, 7-3-1, Hongo, Bunkyo-ku, Tokyo, 113-8656, JAPAN

E-mail: iriguchi.hiroki@jxtg.com

Abstract

To establish guidelines for the development of high activity hydrodesulfurization catalysts, first-principles calculations were carried out with Al_2O_3 as the catalyst support and MoS_2 as the supported metal species, focusing on the TiO_2 coating and support-metal interaction. MoS_2 clusters were modeled on the surfaces of Al_2O_3 and TiO_2 and the adsorption energy was calculated. TiO_2 showed stronger interaction with MoS_2 than Al_2O_3. MoS_2 clusters were also modeled on Al_2O_3 supports coated with thin films of TiO_2, and the adsorption energy was calculated. Adsorption energy depended on the number of stacked film layers, suggesting that thin film thickness can be optimized. An amorphous Al_2O_3 support model was created to approach the real surface state and adsorption calculations were performed. The adsorption energy was higher than with a crystalline support, and the presence of sites where more stable adsorption would occur was confirmed [1].

1. Introduction

Development of highly active hydrodesulfurization (HDS) catalysts requires a fundamental understanding of support-metal interactions, because HDS catalytic activity differs greatly depending on the catalyst support used, even with the same metal species (active sites of the desulfurization reaction). For example, the thiophene desulfurization reaction rate over CoMo catalyst was $k_{\mathrm{HDS}} = 7.8\ \mathrm{cm}^3 / \mathrm{g\ cat.\ h}$ using Al_2O_3 support, whereas the rate was $k_{\mathrm{HDS}} = 36.6\ \mathrm{cm}^3 / \mathrm{g\ cat.\ h}$ using TiO_2 support [2]. The active site of the desulfurization reaction is believed to be the edge site of the supported metal species, so the catalytic activity depends on the electronic state of the supported metal species. The electronic state of the metal species can be affected by the type of support, so interactions between the catalyst support and the supported metal are likely. Depending on the synthesis process a catalyst can be categorized as a Type I or

Type II catalyst. Type I and Type II catalysts contain active sites that are present in large numbers in calcined catalysts and non-calcined catalysts, respectively. Type II sites have higher initial activity in the HDS reaction of 4,6-dimethyl dibenzothiophene (4,6-DMDBT) [3], possibly because Type II sites show weak interaction with the catalyst support, which facilitates desulfurization by promoting hydrogenation of 4,6-DMDBT.

Previously, we attempted to develop HDS catalysts by focusing on this support-metal interaction. Specifically, we attempted to control the support-metal interaction by coating the mesoporous SiO_2 catalyst support with a thin film of TiO_2. As a result, we succeeded in improvement of the catalytic activity for desulfurization of 4,6-DMDBT and hydrogenation of naphthalene [4]. Chemical bonds (O-Ti-O) are gradually formed on the surface of the SiO_2 via the coating treatment, and the surface state gradually changes, taking on the properties of bulk TiO_2. Consequently, as the number of coatings increases, the interaction changes and the catalytic activity improves.

Control of support-metal interactions is considered to be a key requirement for improving HDS performance, but these interactions are difficult to measure in experiments. In such cases, molecular simulation methods such as first-principles calculations can be effective [5~7]. This study used first-principles calculations as a method to quantitatively analyze the support-metal interaction and evaluate how different support species affect this interaction. Specifically, adsorption of MoS_2 onto a TiO_2-coated Al_2O_3 support was investigated.

2. Computational details

2.1. Calculation method

All calculations were performed using VASP [8], a first-principles calculation software package based on density functional theory (DFT). The Perdew-Burke-Ernzerhof functional (PBE) [9], which is a type of generalized gradient approximation (GGA), was used for the exchange-correlation functional, and a plane wave with 400 eV cutoff energy was used as a basis function. The electronic properties were calculated using the projector augmented-wave method (PAW) [10,11] as the pseudopotential, and gamma-centered k-point sampling $2\times3\times3$ for the bulk model and $2\times3\times1$ for the slab model. The Hubbard U parameter [12] value was 3.4 eV in the 4d orbital of the Ti atoms [13] for the crystal structure model.

2.2. Structural modeling

The crystal structure and its surface were selected for the surface model with the aim of minimizing the lattice constant difference at the TiO_2-Al_2O_3 interface. Therefore, the calculation models used the (100) plane of the γ structure Al_2O_3 [14] with a 2×1 supercell in the planar direction and (001) plane of anatase TiO_2 with 3×2 supercell in the planar direction. Based on these structures, the deviation in the lattice constant was about 10% for the TiO_2-coated Al_2O_3 model. Both Al_2O_3 and TiO_2, models were also calculated in which only the supported metal species was adsorbed. To replicate the surface, a vacuum layer of about 10 Å was inserted in the periodic boundary structure in the thickness direction. In optimizing the structure of the adsorption structure, the lower two layers of the slab were fixed to replicate the bulk properties. For example, a model of Mo_6S_{12} supported on an Al_2O_3 surface coated with a single layer of TiO_2 is shown in Fig. 1 a. This calculation assumed a pure surface, without considering the surface modification effects such as the OH group. Therefore, only the direct interactions between the Mo or S atoms and the surface were considered.

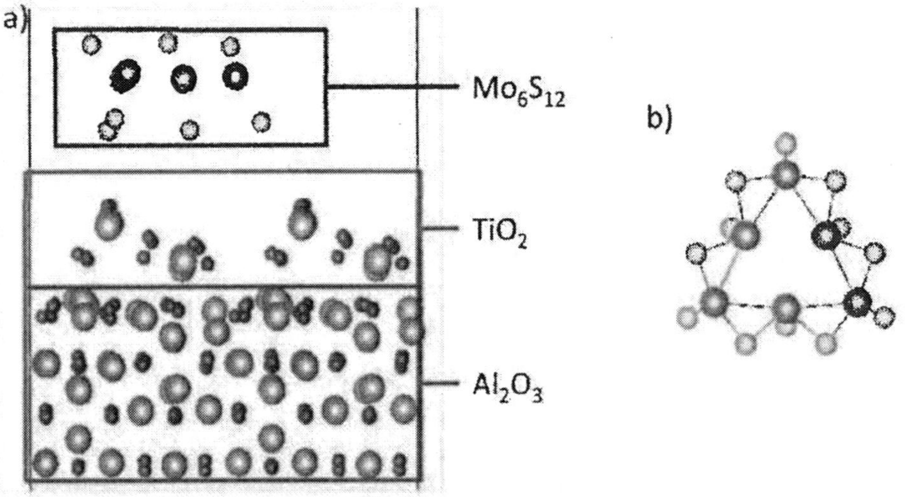

Fig. 1 Calculation model: a) Mo_6S_{12} on Al_2O_3 with one layer of TiO_2.
b) Structure of Mo_6S_{12} cluster (top view).

The structure of the supported metal species cluster was based on MoS_2. Experiments have shown that the structure of the supported metal species on the catalyst surface is hexagonal, but the surface model has limitations within the range of 11.174 Å × 8.413 Å, which is the periodic boundary in the planar direction, and the ratio of Mo : S is 1:2, so the triangular cluster of Mo_6S_{12} (Fig. 1 b) previously proposed as an optimized structure [15] was adopted. This structure has a Mo-S positional relationship similar to that of bulk MoS_2 but is more energetically stable than the cluster structure cut out from the bulk. The molecular structure of the cluster is optimized with the Γ point and sufficiently large periodic boundaries (called the "supercell method"). Structural relaxations of the cluster were also allowed in the surface adsorption calculations.

Support-metal interaction was evaluated as the adsorption energy of the supported metal species to the catalyst support. The adsorption energy $_{ads}$ is defined by Eqn. (1).

$$\Delta_{ads} = {}_{tot} - ({}_{surf} + {}_{mol}) \quad (1)$$

In this equation, ${}_{surf}$ represents the total energy of the catalyst support surface model, ${}_{mol}$ represents the total energy of the supported metal clusters, and ${}_{tot}$ represents the total energy of the model with clusters adsorbed onto the support surface. This energy is an indicator of the extent to which the supported metal species has been thermodynamically stabilized by adsorption on the carrier surface. More negative values indicate more stable the adsorption.

2.3.Amorphous support model

The surface model described in the previous section represents a surface based on a crystal structure. However, the structure of the experimentally produced catalyst support is amorphous, and the effect of this must be considered to compare the experimental findings with the calculated results.

To build an amorphous support structure on which the MoS_2 clusters had adsorbed, we used the *ab initio* molecular dynamics method [16,17] as opposed to classical molecular dynamics, in which the amorphous structure was created by a molecular dynamics method based on the force obtained by calculating the electronic state at each step of the structure. We referred to previous studies [18~20] for the temperature change; 1586 steps were calculated at 5000 K, then, after melting the crystal structure, 1075 steps were calculated at 300 K to relax the structure and bonding states, which can occur at room temperature. Calculations used a time step of 2 fs. After structural relaxation, a TiO_2 monolayer and a vacuum layer were placed on the surface of the structure, which was also relaxed for several steps at 300 K using *ab initio* MD calculations.

3. Results & Discussion

3.1.Adsorption onto Al_2O_3 or TiO_2

Fig. 2 shows a structurally optimized model for the adsorption of Mo_6S_{12} onto Al_2O_3 and TiO_2. The energy of adsorption onto Al_2O_3 was −0.009 eV, which is almost zero considering the calculation accuracy and Mo_6S_{12} was not stabilized by adsorption. In contrast, the energy of adsorption onto TiO_2 was −0.15 eV, which indicates interaction between physical adsorption levels. This ranking of the interaction is consistent with that seen in previous studies [5,6]. In addition, both models confirmed the formation of hybrid orbitals involving the MoS_2 orbitals and the 2p orbitals of the surface O atoms nearby Fermi level, but no trajectory overlap was observed (Fig. 3). This result also suggests that adsorption in this calculation model is physical adsorption, which is considered to be a feature of Type II catalysts.

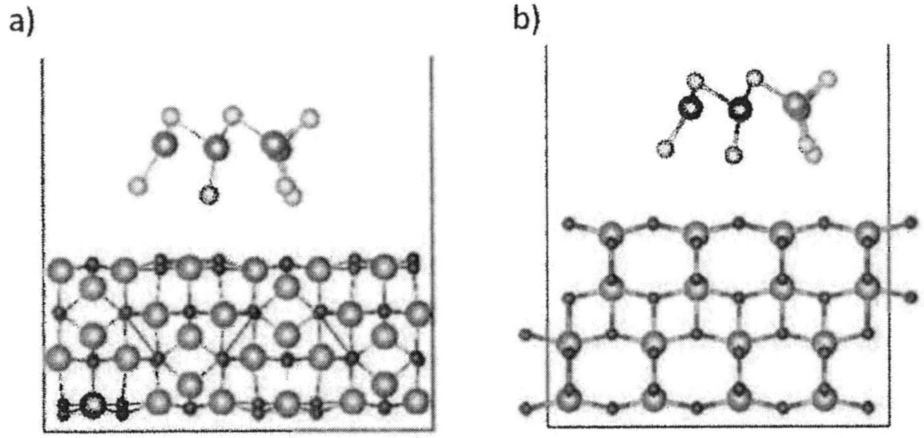

Fig. 2 Optimized structure of Mo_6S_{12} adsorption on a) Al_2O_3 and b) TiO_2

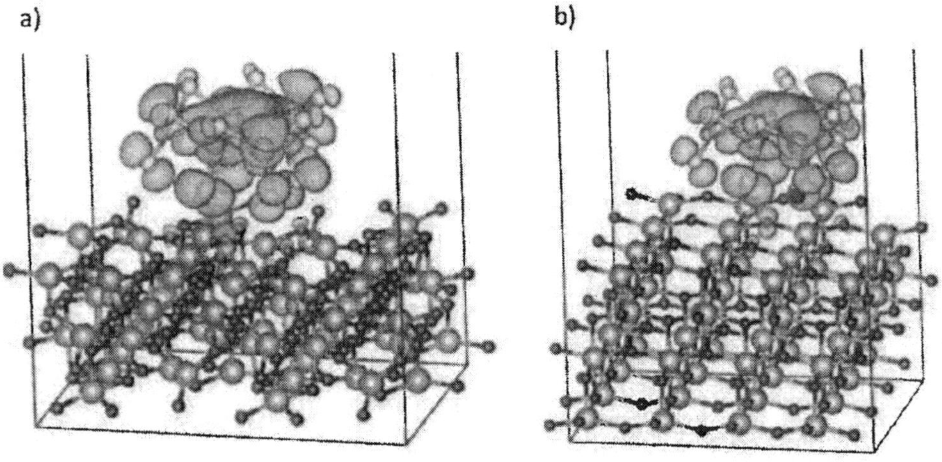

Fig. 3 Molecular orbital of a) Mo_6S_{12}/Al_2O_3 and b) Mo_6S_{12}/TiO_2
The energy of each state is a) +0.26 eV and b) +0.29 eV vs Fermi energy

3.2.Adsorption onto TiO_2-coated Al_2O_3

The energy of adsorption onto Al_2O_3 supports coated with one layer or two layers of TiO_2 is shown as a potential energy surface (PES) in Fig. 4. To reduce the computational load, the most stable adsorption distance and energy were estimated by drawing the PES using single point calculations with different carrier-cluster distances, without optimizing the structure. In the single layer stacking model, the adsorption energy of −0.104 eV at an adsorption distance of 2.6 Å was taken as the minimum value, and in the two-layer stacking model, the adsorption energy of −0.263 eV at an adsorption distance of 2.4 Å was taken as the minimum. Therefore, the adsorption energy in the two-layer stacked model was higher than that (−0.15 eV) with the TiO_2 support. The support-cluster distance and the adsorption energy calculation results for each model are summarized in Table 1.

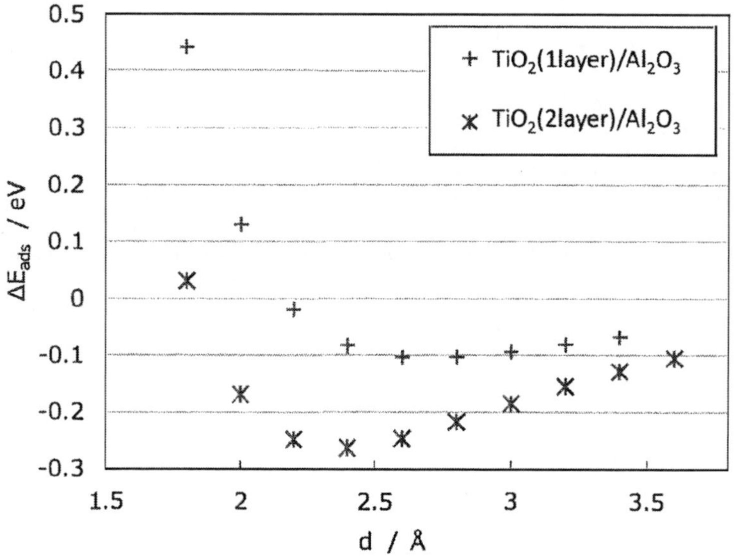

Fig. 4 Potential energy surface (PES) of Mo_6S_{12} adsorption on Al_2O_3 coated with TiO_2

Table 1: Mo_6S_{12} adsorption energy and distance on each support

Support	Adsorption Energy / eV	Distance / Å
Al_2O_3	-0.009	2.7
TiO_2 (1 layer)/Al_2O_3	-0.104	2.6
TiO_2 (2 layer)/Al_2O_3	-0.263	2.4
TiO_2	-0.125	3.2

3.3. Adsorption onto amorphous support

Fig. 5 shows the Mo_6S_{12} adsorption structure and PES for a model in which amorphous Al_2O_3 is coated with a single layer of TiO_2. The minimum adsorption energy was −0.126 eV at a distance d of 2.4 Å. Compared to the energy of adsorption onto crystalline Al_2O_3 coated with a single layer of TiO_2 (−0.104 eV), no dramatic change in interactivity occurred, but the result suggested stronger interaction with amorphous Al_2O_3 than with crystalline Al_2O_3. The uneven distribution of atoms in the amorphous structure may result in locations of maximum interaction.

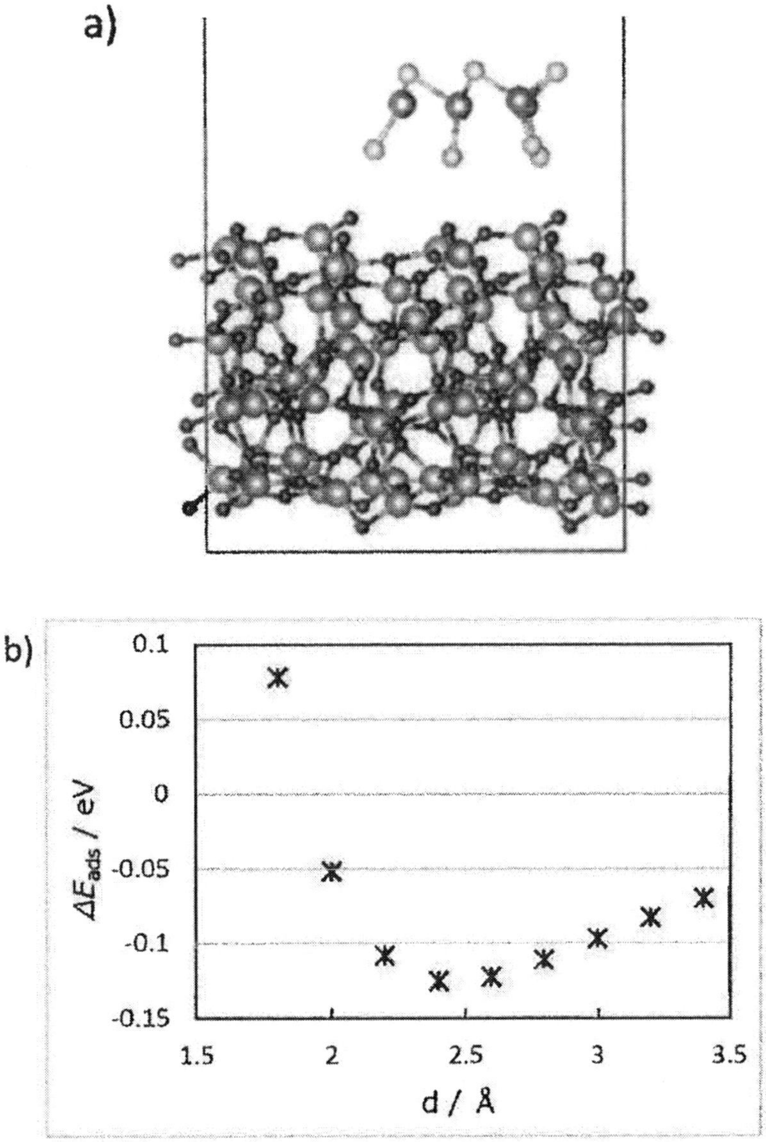

Fig. 5 a) Model of Mo_6S_{12} adsorption on Al_2O_3 coated with one layer of TiO_2 and b) its PES

3.4. Relationships between support-metal interaction and catalytic activity

Finally, we will discuss the interaction and catalytic activity modeled by calculation in this study. Comparison of the differences in support type indicated that TiO_2 had stronger interaction with the supported metal than Al_2O_3, and TiO_2 has higher catalytic activity than Al_2O_3 in experiments. In addition, TiO_2-coated Al_2O_3 also showed stronger interaction with the supported metal than Al_2O_3 and shows higher catalytic activity than Al_2O_3 in experiments [21]. Combining our calculation results about the number of TiO_2 coating layers with previous experiments [4,21] indicates that catalytic activity increased, and interaction was stronger as the number of coating layers increased to optimal thickness. The present models demonstrated the support-metal interaction with positive correlation with catalytic activity. This finding seems to contradict the conventional view that HDS activity increases with weaker interaction. However, as described above, the present model can be interpreted as a Type II catalyst because the support atoms and supported metal atoms are not bound via O, and the adsorption energy is typical of physical adsorption. Therefore, the present model suggests that stronger support-metal interaction will imply larger area of improved catalytic activity in Type II catalysts. These findings indicate that weaker interaction does no imply higher catalytic activity but that control of the strength of the interaction is critical. In this study, only positive correlation was found between interaction and activity, but a volcano plot could be constructed by repeating the calculations with different supports, so that we should be able to identify the value of the optimal interaction.

4. Conclusions

In this study, the interaction between Mo_6S_{12} clusters and various supports consisting of Al_2O_3, TiO_2, and TiO_2-coated Al_2O_3 was investigated by means of first-principles calculations. Adsorption calculations with Al_2O_3 and TiO_2 supports showed that Mo_6S_{12} is stabilized more by adsorption onto TiO_2 than onto Al_2O_3, suggesting that TiO_2 interacts strongly with the active metal species. The interaction became stronger as the number of coating layers increased on TiO_2-coated Al_2O_3 supports. Since the maximum value of this interaction is larger than that for the simple TiO_2 support, interaction will be maximized at the optimum coating thickness. Finally, cluster adsorption was calculated for an amorphous support structure. Interaction was stronger than that with a crystalline support, suggesting that more stable adsorption sites had formed as a result of amorphization.

In the range of the present calculations, the support-metal interaction and the catalytic activity showed a positive correlation. Similar calculations with other supports and supported metal species may provide deeper insight into this correlation, possibly allowing the use of simulations to develop the optimum catalyst for any reaction.

References

[1] Iriguchi, H., Uratani, H., Kudo. T., Seki, H., Yamashita, K., *J. Jpn. Petrol. Inst.*, **61**, 288 (2018)

[2] Borque, M.P., Lopez-Agudo, A., Olguin, E., Vrinat, M., Cedeno, L., Ramirez, *J., Appl. Catal. A: General*, **180**, 53 (1999).

[3] Berhault, G., Perez De la Rosa, M., Mehta, A., Yacaman, M.J., Chianelli, R.R., *Appl. Catal. A,* **345**, 80 (2008).

[4] Satake, H., Okamoto, M., Sakayori, Y., Furudera, S., Kubota, T., Seki, H., *J. Jpn. Inst. Energy*, **96**, 114 (2017).

[5] Joshi, Y.V., Ghosh, P., Daage, M., Delgass, W.N., *J. Catal.*, **257**, 71 (2008).

[6] Arrouvel, C., Breysse, M., Toulhoat, H., Raybaud, P., *J. Catal.*, **232**, 161 (2005).

[7] Weber, T., Rob van Veen, J.A., *Catal. Today*, **130**, 170 (2008).

[8] Hafner, J., *J. Comput. Chem.*, **29**, 2044 (2008).

[9] Perdew, J. P., Burke, K.., Ernzerhof, M., *Phys. Rev. Lett.*, **77**, 3865 (1996).

[10] Kresse, G., Furthmüller, J., *Phys. Rev. B: Condens. Matter Mater. Phys.*, **54**, 11169 (1996).

[11] Blöchl, P. E., *Phys. Rev. B: Condens. Matter Mater. Phys.*, **50**, 17953 (1994).

[12] Dudarev, S.L., Botton, G.A., Savrasov, S.Y., Humphreys, C.J., Sutton, A.P., *Phys. Rev. B*, **57**, 1505 (1998).

[13] Kamisaka, H., Mizuguchi, N., Yamashita, K., *J. Mater. Sci.*, **47**, 7522 (2012).

[14] Digne, M., Saulet, P., Euzen, P., Toulhoat, H., *J. Catal.*, **226**, 54 (2004).

[15] Murugan, P., Kumar, V., Kawazoe, Y., Ota, N., *Phys. Rev. A*, **71**, 063203 (2005)

[16] Kresse, G., Hafner, J., *Phys. Rev. B: Condens. Matter Mater. Phys.*, **47**, 558 (1993).

[17] Kresse, G., Hafner, J., *Phys. Rev. B: Condens. Matter Mater. Phys.*, **49**, 14251 (1994).

[18] Adiga, S.P., Zapol, P., Curtiss, L.A., *Phys. Rev. B.*, **74**, 064204 (2006).

[19] Gutiérrez, G., Johansson, B., *Phys. Rev. B.*, **65**, 104202 (2002).

[20] Matsui, M., *GOLDSCHMIDT CONFERENCE EDINBURG*, **58A**, 571 (1994).

[21] Segawa K., Takahashi, K., Satoh, S., *Catal. Today*, **63**, 123 (2000).

Application of inferential models to crude distillation process: integrating laboratory feedback in closed-loop quality control

Rashid M. Ansari, PhD
Engineering Specialist - Refining Development
Oil Upgrading R&D Division
Research & Development Center, Saudi Aramco

ABSTRACT

The inferential models, based on fast and continuously available temperature, pressure and flow measurements reduce the negative impact of sample intervals and time delays with minimum compromise on accuracy. The resulting continuous and fast response keeps the product qualities on specifications and minimizes the quality giveaway.

A wide range of inferential models were developed on crude fractionation unit for various product qualities and implemented these models in real-time to form a closed-loop quality control using laboratory feedback mechanism. The model is updated using laboratory results as a basis to update the predicted values and a first-order filter to compensate for the noise in the flow measurements. The average deviation from the laboratory results is only 2 °C, and considering the repeatability of ± 4 °C in Naphtha FBP laboratory results, this prediction can be considered very accurate.

The development of 85% combined diesel inferential model was an innovative approach to inferential modelling by incorporating the blending concept in inferential model development which combined many streams from the two units to predict the quality of combined diesel and controlled that to required specification. The savings from these models are significant by eliminating the need of expensive analyzers and exclusively using the laboratory mechanism in closed-loop quality control system.

This article highlights the development aspects of correlations, use of inferential model in open and closed loop environment, integration of laboratory update mechanism, operator's acceptance and the benefits of inferential modelling.

Catalytic Dealkylation of Alkyl Polycyclic Aromatic Hydrocarbons towards Innovation of Petroleum Refinery

Naonobu Katada and Satoshi Suganuma
Center for Research on Green Sustainable Chemistry, Tottori University, 4-101 Koyama-cho Minami, Tottori 680-8552, Japan
katada@chem.tottori-u.ac.jp

Introduction

Among fractions produced by petroleum refinery [1], polycyclic aromatic hydrocarbons such as naphthalene, phenanthrene, pyrene and their derivatives have little usage. In contrast, the C6-10 fraction useful as gasoline has been preferred to contain high concentration of monocyclic aromatic hydrocarbons, mainly BTX [benzene, toluene (methylbenzene) and xylenes (dimethylbenzene isomers)] due to their high octane values [2], whereas the fraction of hydrocarbons larger than C10 applicable to diesel fuel has been demanded to be enriched with alkanes possessing high cetane values [3]. The BTX is also the raw material of some of the polymers produced and consumed in the largest scale [4]. As well as the fluid catalytic cracking (FCC) maximizing the C5-10 hydrocarbons enriched with BTX [5], in these decades, upgrading techniques based on hydrocracking / hydroprocessing have much been advanced [6-9], because suppression of the aromatic contents of gasoline and diesel fuel for the environmental protection has been desired [10-12] as well as the removal of hetero element (N, S and metals) compounds [13-16], and increase of the demand for diesel fuel [17,18] caused surplus of the aromatic hydrocarbons including BTX in some extent [10,19]. In the hydrocracking / hydrotreating, the aromatic hydrocarbons are hydrogenated into the compounds with less aromatic rings [9,20-22], as postulated by the direction from the top to the bottom of Figure 1 (A). The cracking makes the smaller molecules from the larger ones [21-23], i.e., from the right to the left. Combining these reactions, the amount of gasoline, naphtha and diesel fractions are relatively increased [9,24].

Nowadays, importance of the carbon dioxide control and decrease of the growing rate of demand for fuels including gasoline and diesel fuel have been recognized, and therefore we can count disadvantages of conventional hydrocracking in very recent or future situation as the following (a)-(d).

(a) It is difficult to increase the yields of BTX and >C10 alkanes only, causing more consumption of the carbon source and finally the carbon dioxide emission to keep the sufficient supply of BTX and >C10 alkanes.

(b) The practical feed of hydrocracking, e.g., LCO (light cycle oil), with low hydrogen / carbon ratio results in serious catalyst deactivation due to coke formation [25]. Remarkable improvement of reaction conditions [26] and catalyst [27,28] to overcome this problem should be mentioned, but it is necessary to consume large amount of hydrogen for the reaction of a heavy carbon source [9,17,25,29], and it can be one factor of carbon dioxide emission through hydrogen production.

(c) Strong hydrogenation conditions give low yield of BTX in naphtha fraction. The short term prediction emphasizes surplus of total aromatic hydrocarbons including BTX [10], but again the increase of BTX yield will be demanded, because it is the raw material of polymers [4], while all of other types of hydrocarbons will become gradually surplus due to the decrease of demand for fuels [30]. Therefore the low yield of BTX must be the problem in near future.

(d) The cracking of hydrogenated intermediates, i.e., alkanes and cycloalkanes, mainly proceeds via dealkylation of alkyl aromatic intermediates formed by hydrogenation-dehydrogenation processes [21,31], and it generally forms too small alkanes (<C5) at high conversion [24].

An alternative can be drawn in Figure 1 (B). It is predicted that dealkylation of alkyl polycyclic aromatic hydrocarbons (hereafter APAH) via carbenium cation is readily catalyzed by Brønsted acid like cracking of cumene (2-phenylpropane) [32,33]. Hydrogenation of the alkene into alkane should be also easy [19] and requires only one H_2 molecule. The formed alkanes and aro-matic compounds with no alkyl branches can be separated by such methods as solvent extraction [34] to give the alkanes whose chain lengths are kept, maximizing the possibility for use as the diesel fuel.

On the other hand, aromatic compounds without alkyl branch, i.e., benzene, naphthalene, anthracene, phenanthrene and pyrene etc., will be obtained. Benzene is directly a chemical resource [4]. Partial hydrogenation of polycyclic aromatic ring into tetralin derivative, e.g., naphthalene into tetralin [12,22, 35 - 37], is possible on such a catalyst as zeolite-supported molybdenum compound [21,24,35,36]. Ring-opening (dealkylation) of tetralin has been known to proceed over the zeolites with relatively large micropores [6,21,22,35,38]. Thus, this alternative process can maximize the amount of benzene derivatives and long chain alkanes with the minimum consumption of hydrogen. The alkanes with larger than the required molecular weight range is easily converted into smaller ones by conventional FCC (fluid catalytic cracking); it is noteworthy that the feed in this case has no aromatics. Even after the demand for diesel and then gasoline as a fuel will decreased, the alkanes will be able to convert into benzene derivatives by repeating the FCC and dealkylation. The final goal of our attempt is hence to realize the complete innovation of petroleum refinery like Figure 2 to get the maximum conversion of crude to chemicals.

In order to realize it, it is necessary to develop the following processes (A)-(D).

(A) Dealkylation of APAH
(B) Separation of aromatic compounds and alkanes
(C) Partial hydrogenation of naphthalene, anthracene, phenanthrene and pyrene etc.
(D) Ring-opening of tetralin derivatives.

This paper reports an attempt to develop (A).

Dealkylation of a small alkyl aromatic compound such as cumene is known to be an easy reaction catalyzed by a weak Brønsted acid [32,33]. Related reactions, alkylation and transalkylation of alkylbenzene [39-42], dealkylation of long chain alkylbenzene [43] and disproportionation of alkylnaphthalenes [44] have been studied mainly on zeolites. Typical solid acids such as zeolite and amorphous silica-alumina catalyze not only the dealkylation shown by below (1) but also side reactions such (2) and (3), as stated later. In addition, it has been known that the reaction type (2) preferentially occurred compared to (1) in the non-catalytic reaction [45]. We examined the performances of several catalysts to find a catalyst to selectively enhance the reaction rate of desired reaction, dealkylation (1) and successive hydrogenation (4).

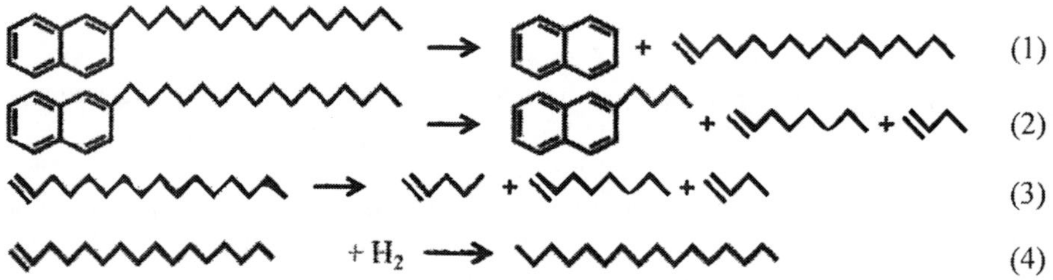

We here examined aluminosilicates which possessed intrinsically high thermal stability. In order to represent the APAH compounds in typical VGO, reagents of alkylnaphthalene was used as model reactants on the first stage of study, and then a practical sample of VGO was examined. In the present study, the components of aromatic and alkyl parts were quantified with an aid of two-dimensional GC (2D-GC) technique [46,47] which has recently been applied to the analysis of heavy oils [48].

In this study, we examined a silica monolayer loaded on alumina as a candidate of catalyst. Chemical vapor deposition (CVD) of Si precursor on the surfaces of oxides with weakly or moderate basic nature such as alumina and zirconia have been tested to form new catalysts or catalyst supports by several authors [49-58]. Among them, the silica layer on alumina has been found to have catalytic activity for Brønsted acid-catalyzed reactions [50,59] and very high thermal stability [54,57,60]. The influences of precursor of silica, CVD temperature and calculation temperature have been studied extensively as follows. Using tetramethoxysilane [$Si(OCH_3)_4$] at 523-673 K gives a mono layer con-sisting of a two-dimensional network of Si-O-Si with a small concentration of SiOH, while using $Si(CH_3)_n(OCH_3)_{4-n}$ (n = 1-3) or CVD [61] of tetramethoxysilane at <523 K [62] gives the isolated species. The CVD at >673 K gives heterogeneous thickness of the layer [62]. The Si-O-Al bond was formed during the CVD process, and the remained -OCH_3 group was removed by calcination of the sample at >673 K, as observed by in-situ IR experiments [59] and weight monitoring [62]. The calcination at 673-1173 K gives principally stable properties where the Si concentration was 8-12 nm^{-2} [60]. Based on these facts, we selected tetramethoxysilane as the precursor of silica, 593 K as the CVD temperature and 673 K as the calcination temperature in this study.

In addition, a method of ammonia IRMS (infrared-mass spectroscopy) - TPD (temperature-programmed desorption) was applied to show the acidic property in this study. In this method, absolute quantification of desorbed ammonia is carried out with MS while the temperature dependences of desorption of ammonia from Brønsted and Lewis acid sites are recorded by IR. Combined analysis based on a curve fitting method gives TPD profiles of Brønsted and Lewis acid sites, and then the number and distribution of enthalpy of ammonia desorption as an index of acid strength of each of Brønsted and Lewis acid sites can be obtained [63,64]. In 1990s when the silica monolayer was invented, we had no method for precise measurements of number and strength of each of Brønsted and Lewis acid sites. The discussion was carried out mainly based on the test reactions [62]. In the last decade, we have extensively studied the theory and experimental method of ammonia TPD. This is also the first report of the application of ammonia IRMS-TPD to the silica monolayer.

Experimental

Catalyst preparation

Table 1 shows catalysts employed in this study.

Thin silica layers were prepared by means of a CVD method [59] on alumina JRC-ALO-6, which had been supplied by Catalysis Society of Japan as a reference catalyst. The deposition was performed in a vacuum system using $Si(OCH_3)_4$ (tetramethoxysilane) as the precursor of silica. After evacuation of the support alumina at 673 K for 2 h, vapor of tetramethoxysilane was admitted to contact with alumina at 593 K, and the resultant increase of weight was monitored using a quartz spring balance. The degree of vacuum was ca. 10^{-3} Torr (1 Torr = 133.3 Pa), and the vapor pressure of alkoxide was kept at ca. 2.5 Torr by chilling the reservoir with an ice bath. After the deposition until the approximately desired weight gain was observed, the sample was calcined in oxygen (200 Torr) at 673 K for >4 h. The amount of deposited Si was calculated from the weight increase with assuming that the formed material was SiO_2.

In addition, three samples of amorphous silica-alumina catalysts industrially available, N631-L (JGC Catalysts and Chemicals Ltd.), reference catalysts JRC-SAH-1 and JRC-SAL-2 (Catalysis Society of Japan, hereafter SAH-1 and SAL-2, respectively) were tested. An Na-Y zeolite 320NAA with FAU structure (Si/Al_2 = 5.9 and [Al] = 4.4 mol kg^{-1}) was supplied by Tosoh Corp., ion-exchanged into NH_4-Y, steamed at 823 K, treated with an NH_4NO_3 aqueous solution at 353 K and dried at 373 K into NH_4-form. It was converted into H-form during the pretreatment of reaction (in-situ preparation). Hereafter it is termed N-USY [65]. Reference catalysts JRC-Z-HB25 (1) with *BEA structure (Si/Al_2 = 25 and [Al] = 1.2 mol kg^{-1}) and JRC-HM20 with MOR structure (Si/Al_2 = 20 and [Al] = 1.5 mol kg^{-1}) distributed by Catalysis Society of Japan were also employed. In addition, a Na-ZSM-5 zeolite HSZ820NAA with MFI structure (Si/Al_2 = 24 and [Al] = 1.3 mol kg^{-1}) was supplied by Tosoh Corp., ion-exchanged with an NH_4NO_3 aqueous solution at 353 K and dried at 373 K into NH_4-form.

Ammonium molybdate [$(NH_4)_6Mo_7O_{24}$] was loaded on alumina JRC-ALO-6 in an aqueous solution with keeping pH at 2 by adding nitric acid, filtered and calcined at 773 K in an oxygen flow to prepare MoO_3/Al_2O_3. The surface concentration of Mo was 5 atoms nm^{-2}, which was determined by ICP-ES (inductively coupled plasma emission spectroscopy, Rigaku CIROS-CCD).

Structural analysis

The nitrogen adsorption isotherm was measured at 77 K using a BELSorp-max equipment (Microtrac-BEL) after pretreatment at 573 K. The surface area was calculated by BET (Brunauer-Emmett-Teller) equation [66], and the pore diameter distribution was analyzed from the desorption branch using a BJH (Barrett-Joyner-Halenda) method [67]. Contribution of the micropore wall to the surface area was evaluated based on a t-plot method [68].

The amount of chemisorbed benzoate anion was measured by a BAT (benzaldehyde-ammonia titration) method [69]. About 40 mg of the sample was packed in a Pyrex tube (4 mm i.d.) and pretreated in a helium flow (27 μmol s^{-1}, atmospheric pressure) at 673 K for 1 h. Benzaldehyde (9.8 μmol) was injected into the sample bed at 523 K in the helium flow, and the eluted aldehyde was monitored by a GC with a silicone SE-30 packed column and an FID (flame ionization detector). The injection of aldehyde was repeated several times until the saturation of adsorption was observed, and then ammonia (0.41 mmol) was repeatedly supplied at 673 K. The formed benzonitrile was quantified by the GC, and the coverage was calculated on the assumption that the benzoate anion was adsorbed only on the alumina exposed surface but not on the silica-covered surface [59].

The ammonia IRMS-TPD method was applied to measure the number and strength distribution of each of Brønsted and Lewis acid sites [63]. A self-supporting disc (1cm diameter), molded from the sample by compression, was held by a set of metal rings to fix in a cell of Microtrac-BEL IRMS-TPD analyzer and pretreated at 773 K in oxygen. The IR spectra were collected with heating the sample in a helium flow (68 μmol s^{-1}, 6 kPa) at 2 K min^{-1}. Then ammonia was adsorbed at 333 K, and the sample was again heated under the same conditions. The IR spectra and MS response were recorded. The ammonia TPD profile of each of Brønsted and Lewis acid site was analyzed according to our previous study [63]. The number of acid sites was calculated from the peak intensity of the TPD, and the distribution of enthalpy of ammonia desorption (so-called adsorption heat) was analyzed by the curve fitting method [70]. In these experiments, the ammonia adsorption in the ammonia IRMS-TPD analysis was carried out at a relatively low temperature 333 K to keep ammonia on the acid sites with ca. >110 kJ mol^{-1} of the enthalpy of ammonia desorption. We have confirmed that no molecular water was present in these conditions, and that ammonia fully covered the acidic OH groups even in the micropores by the monitoring IR spectrum.

Reaction tests

Dealkylation of cumene (2-phenylpropane) was carried out by means of a pulse method, widely utilized for quick test for solid catalysts, to study the activity for the reaction similar to the target reaction, but the molecular size of reactant was small. The sample was placed in a Pyrex tube (4 mm i.d.). The amount of sample was adjusted to keep the differential conditions (conversion less than 5 %). It was pretreated in a helium flow (41 μmol s^{-1}, atmospheric pressure) at 773 K for 1 h. Cumene (29 μmol) was then injected at 623 K in the helium flow from the septum at the inlet of reactor, and the eluted products were analyzed by a GC connected to the outlet of the reactor. The reaction rate is shown by the amount of consumed cumene divided by the weight of catalyst.

The dealkylation of alkylnaphthalene was carried out as follows. Two types of model reagents of alkylnaphthalene were employed as the reactants. A mixture of hexadecylnaphthalene and octadecylnaphthalene (61 : 39 in molar ratio, consisting of isomer mixture of hexadecylnaphthalene with one alkyl branch whose carbon number was 16 and isomer mixture of octadecylnaphthalene with one alkyl branch whose carbon number was 18) was purchased from Wako Pure Chemical Industries Ltd. Hereafter it is termed C16, 18 AN (alkylnaphthanene) reagent. On the other hand, an isomer mixture of hexadecylnaphthalene with one alkyl branch, whose carbon number was 16, was purchased from Exxon Mobil Corp. as a commercial sample Synesstic 5. It is termed C16 AN reagent in the following sections.

The catalyst was packed in a stainless steel tube (4 mm i.d.). In typical conditions, the catalyst (solid acid catalyst 50 mg + MoO_3/Al_2O_3 20 mg) was pretreated in a flow of hydrogen (0.12 mol h^{-1}) under atmospheric pressure at 773 K for 1 h. Then, the reactant was fed at 0.96 g h^{-1} with an H_2 flow (0.12 mol h^{-1}) through the catalyst bed at 573 K. In some cases, helium was fed in place of H_2. The total pressure in the tube was kept at 1 MPa using a back pressure valve connected to the outlet. The reaction conditions will be clearly shown by the liquid hourly space velocity (LHSV) defined as [rate of feeding reactant AN (in the standard cases, 0.96 g h^{-1}) / weight of solid acid catalyst employed (in the standard cases, 0.05 g)]. Mesitylene (1,3,5-trimethylbenzene) was fed at g h^{-1} between the catalyst bed and the back pressure valve in order to dissolve and flow solid compounds.

The formed material was trapped in a glass tube chilled with an ice bath. The products were analyzed using a 2D-GC (Agilent 7890) with two capillary columns (0.25 μm thickness DB-5MS in length 30 m and i.d. 0.250 mm, and 0.15 μm thickness DB-17HT in length 5 m and i.d. 0.250 mm). An inner standard method was adopted with tetraethylene glycol dimethyl ether as the standard material added after collecting the products. However, mainly because the tailing of GC peaks, the quantification of products contained uncertainties. The components were calculated based on the relative intensities of GC peaks.

Results

Physicochemical properties of catalysts

From the adsorption isotherm of nitrogen at 77 K, the surface area was analyzed based on the BET method [66]. The BET surface area divided by the weight of alumina as a function of the amount of deposited silica, which is shown as the surface concentration (Si atoms nm^{-2}). The change of surface area was little, in consistent with that the silica formed a very thin layer which did not affect the morphology of the particles.

The BAT method showed that the coverage increased with the deposition of silica. It was in proportion to the amount of silica from 0 to 12 Si nm^{-2}. At 12 Si nm^{-2}, the coverage reached to >90% and kept with further deposition. Because the surface concentration of Al atoms on alumina is believed to be ca. 12 nm^{-2}, it is suggested that the monolayer of silica with 1 : 1 bonding such as Al-O-Si covered almost completely the surface at 12 Si nm^{-2}, as already reported [59].

The area of external surface [the surfaces other than the micropore (diameter < 2 nm) wall, therefore, including the walls of meso- (2 to 50 nm) and macropores (> 50 nm)] was calculated by means of the t-plot analysis [68]. The difference between the total surface area (BET) and external surface area is assumed to give the area of micropore wall. The area of mesopore wall is estimated by BJH method [67]. The micropore wall area was negligible on alumina, the silica monolayer loaded on alumina and the amorphous silica-aluminas; the t-plot itself shows the absence of micropore condensation by the adsorption capacity proportional to the estimated thickness of adsorbed material in 0-0.5 nm of the thickness range (Figure S3). On these catalysts, the area of mesopore wall was close to the total surface area, indicating that the mesopores predominantly contributed to form the reaction fields. In contrast, the micropore wall area was close to the total surface area on the zeolites.

The mean pore diameter of alumina-supported silica monolayer was 25 nm, as detemined by the BJH method. The amorphous silica-aluminas had far smaller mesopores in the following order: SAH-1 (6 nm) > SAL-2 (5 nm) > N631-L (4 nm). The micropore diameters of zeolites have been known to be <0.8 nm [71], clearly smaller than these values.

The number and strength distribution of Brønsted and Lewis acid sites were analyzed by ammonia IRMS-TPD method. Brønsted acidity was generated by the deposition of silica on alumina. The maximum of number of Brønsted acid site was observed at around 11 Si nm^{-2}, where the monolayer of silica almost completely covered the surface. It suggests that the Brønsted acid site was generated on the monolayer.

The enthalpy of ammonia desorption (so-called adsorption heat) was determined by analysis of the ammonia TPD spectrum to provide a measure of acid strength. We have reported that the enthalpy of Brønsted acid site generated by the framework Al in zeolite had a sharp distribution and dependent on the framework type like MOR (150-160 kJ mol^{-1}) > MFI (130-145 kJ mol^{-1}) > *BEA (125-135 kJ mol^{-1}) > FAU (105-115 kJ mol^{-1}) [64,72]. The enthalpy of USY zeolite with FAU structure was however modified by the extra-framework Al species, and that of NH_4NO_3-treated USY (N-USY) was ca. 150 kJ mol^{-1} [65]. Thus various acid strengths were observed on the zeolites. It is believed that the microstructure around the acid site, e.g., interatomic distances and angles, varies the acid strength, and therefore the acid strength was dependent on the framework topology [73]. In contrast, broad distributions of ammonia desorption enthalpy on the alumina-supported silica monolayer and the amorphous silica-aluminas were found. It ranged from 100 to 170 kJ mol^{-1}. It is suggested that the structure around acid site was widely varied on these materials due to the lack of crystallinity.

Catalytic performance in reactions of AN reagents

Table 2 compares the catalytic activities for the dealkylation of cumene on the alumina-supported silica monolayer and amorphous silica-aluminas. Generally the activity was low on the silica monolayer, and then SAH-1 < SAL-2 < N631-L. It is noteworthy that the activity of zeolite for this reaction was quite high, at least 20 mol g^{-1} on ZSM-5 and Y, and it was impossible to keep the differential conditions and therefore impossible to quantify the reaction rate in the same conditions.

In these conditions, the dealklylation of alkylnaphthalene into naphthalene and alkenes followed by the hydrogenation of alkenes into alkanes is considered to proceed. The reaction rate slightly decreased with time on stream, but the extent of deactivation was small in the experimental range. Hereafter, the catalytic performance will be shown by the averaged values of conversion and yield for ca. 3 - 8 h of the time on stream where stable activity and selectivity were observed.

Figure 3 shows the catalytic activity for this reaction as a function of the amount of deposited silica on alumina. The alumina support showed little activity. The activity was created by the deposition of silica and showed the maximum at around 10 Si nm^{-2}, where the monolayer of silica approximately covered the surface. The further deposition decreased the activity. A similar trend was observed in a different set of reaction conditions. Although a mixture of silica layers and MoO_3/Al_2O_3 was employed for the above reactions, it has been found that the role of MoO_3/Al_2O_3 was small. The selectivity on the silica-deposited samples was close to 100 %. The distribution of C16 and 18 among the formed alkanes was also close to 100 %. The selectivity was kept high in 573 - 673 K but decreased at higher temperatures.

Experiments using a practical sample of VGO showed that the reaction was prohibited by basic materials in practical petroleum. After the removal of basic components by treatments with acidic ion exchange resin, the dealkylation of alkylnaphthalene and alkylpyrene proceeded in VGO.

Figure 4 compares the activities and selectivities of various aluminosilicate solid acid catalysts. Both of the activity and selectivity of amorphous silica-aluminas were slightly lower than those of silica monolayer, resulting in the lower yield of desired products. The yield was in the order of SAH-1 > SAL-2 > N631-L; the difference was small. The zeolite β showed the low selectivity due to the formation of both of alkylnaphthalenes with alkyl groups shorter than C16 and alkanes smaller than C16. The N-USY zeolite showed the high selectivity but low activity. The ZSM-5 zeolite showed little activity and, among the formed alkanes, only alkanes shorter than C16 were detected. The activity for the dealkylation was thus found to be in the order of silica monolayer > SAH-1 > SAL-1 > N631-L > N-USY > β > Y > Mordenite > ZSM-5, as shown by the yield of desired products.

Thermal stability

Figure 5 shows changes in BET surface area and activity for the dealkylation of alkylnaphthalene on the alumina-supported silica monolayer and an amorphous silica-alumina (N631-L) by high temperature calcination. The fresh catalyst was calcined at each temperature and then tested in the reaction, and the reaction rate per unit weight in differential conditions (low conversion) was evaluated here in order to compare the intrinsic activity. The surface area and activity of the silica monolayer were kept constant up to 973 K and gradually decreased at higher temperature. The activity was slightly decreased at 1173 K and still observed at 1173 K. The surface area and activity of amorphous silica-alumina was more quickly decreased in 973-1173 K and completely lost at 1373 K. We have reported that the silica monolayer showed the high surface area and Brønsted acidity (activity for double bond isomerization of 1-butene) after high temperature calcination [60], and that the high thermal stability was due to the low concentration of OH groups [74]. It is predicted that high temperature calcination will be necessary for regeneration of the catalyst deactivated by strong adsorption of heavy compounds in practical processes. The durability of silica monolayer catalyst against the high temperature treatment is thus-demonstrated.

Discussion

Comparison of aluminosilicate solid acid catalysts

The catalytic activities for both reactions of cumene and alkylnaphthalene were generated by the deposition of silica on alumina, whereas the pure alumina showed negligible activity. The activity showed the maximum at around 10 Si nm^{-2} where the silica monolayer covered the surface, and further deposition decreased the activity. This was not related with the total acid amount, but the amount of Brønsted acid sites measured by the ammonia IRMS-TPD method well explained these changes. The Brønsted acid amount was negligible on the pure alumina, but increased with the deposition of silica up to ca. 12 Si nm^{-2}. Further deposition decreased the Brønsted acid sites. It indicates that the Brønsted acid site generated on the silica monolayer was the active species for the dealkylation of alkyl aromatic compounds. It should be noted that the textural and porous property of alumina is maintained during the CVD of silica as confirmed by the stable BET surface area and electron microscopy in our previous paper [75]. It can be concluded that the activity was dependent on the Brønsted acid amount where the textural and porous property was equivalent.

The influence of Brønsted acid strength on the activity for dealkylation of alkylnaphthalene was observed in a few examples. The Brønsted acid strength of N-USY was higher than that of Y [65], whereas the framework and porous structure were same to give the same steric influence. The N-USY zeolite showed higher activity than Y (Figure 4), suggesting that the relatively stronger Brønsted acid site had higher activity.

However, no clear dependence of the activity on Brønsted acid strength and its distribution was found in comparison of catalytic performances among catalysts with different structures. The activity for the desired reaction, dealkylation of alkylnaphthalene, was roughly in the order of silica monolayer > SAH-1 > SAL-2 > N631-L > zeolites (Figure 4). The Brønsted acid strength of silica monolayer and amorphous silica-alumina showed broad distributions, while the zeolites had different strengths with relatively narrow distributions as stated in the previous section. The activity was not related with the Brønsted acid strength.

The activity on MFI zeolite was obviously low, and therefore the activity for dealkylation of alkylnaphthalene can be shown as silica monolayer > SAH-1 > SAL-2 > N631-L > zeolites with 12-rings (FAU, *BEA and MOR) > zeolite with 10-ring (MFI) (Figure 4). This was in agreement with the order of pore size. It is suggested that the steric hindrance was predominant in the kinetics of this reaction, and the silica monolayer showed the relatively high activity because of its large pore size; as the pore diameter of silica monolayer was 25 nm, far larger than the related molecules, and therefore it is perhaps better to say that the present silica monolayer catalyst had non-porous structure resulting in the high activity.

In conclusion, the activity for dealkylation of alkylnaphthalene was sensitive to the pore size rather than the acid strength. The dealkylation of alkyl aromatic hydrocarbon can be catalyzed even by a weak Brønsted acid [32,33]. In the present study, it was indicated that the presence of Brønsted acid site was necessary. However, it was shown that the influence of pore size was predominant compared to the acidic property, presumably because the molecular size of reactant was large. On some of the amorphous silica-aluminas and zeolites, the by-products, i.e., alkylnaphthalenes with shorter alkyl chains and shorter alkanes, were detected. It implies that, in the narrow pores, the cracking of linear alkyl part proceeded readily compared to the reaction of dealkylation of the same molecule, probably due to the poor accessibility of the base of alkyl branch attaching to the aromatic ring.

The activity for the dealkylation of smaller molecule, cumene, was roughly in the reverse order of the activity of the large molecule, alkylnaphthalene (Table 2). It has not been possible to show what controlled the activity of cumene cracking, one can say at least that the activity for dealkylation of bulky molecule is not related with the activity for the same kind of reaction of smaller molecule. It supports that the activity for the reaction of large alkyl aromatic compound was strongly dependent on the accessibility of molecules.

As shown in the previous section, the followings should be studied for maximizing the yield of monocyclic aromatic compounds, mainly BTX, as well as maximizing the possibility for use of alkyl parts as the diesel fuel.

(A) Dealkylation of APAH
(B) Separation of aromatic compounds and alkanes
(C) Partial hydrogenation of naphthalene, anthracene, phenanthrene and pyrene etc.
(D) Ring-opening of tetralin derivatives.

The present paper demonstrates the possibility of part (A), and the detailed contents have been published [76]. After the publication, we have established the analysis of reaction and required separation of basic compounds in practical heavy oil samples [77]. Besides, (B) - (D) should be also investigated. Among them, the separation (B) is an easy subject, because everything is controlled by equilibrium, and the complete conversion of alkylaromatics will make the separation easy. The partial hydrogenation (C) [21,24,35,36] and the ring-opening (D) [6,21,22,35,38] have been studied to find suitable catalysts, and we have confirmed fundamental possibilities. In other words, we have almost finished the experimental works in the laboratory scale and now the business partner for practical study is waited. The present study will open a way for maximizing the production of values chemicals from petroleum.

Conclusions

A silica monolayer catalyst was prepared by CVD of tetramethoxysilane on alumina. It showed Brønsted acidity and thermal stability. The silica monolayer showed high activity and selectivity for dealkylation of C16, 18 alkylnaphthalene into naphthalene and C16, 18 alkanes. Among aluminosilicates (silica monolayer, amorphous silica-alumina and zeolites), the activity and selectivity of the silica monolayer were relatively high. The activity was observed even after calcination at 973-1173 K. It is suggested that the silica monolayer catalyst had non-porous structure (pore diameter ca. 25 nm) resulting in the high activity. Tests for application to VGO were carried out, showing potential of the silica monolayer catalyst for dealkylation of alkyl polycyclic aromatic hydrocarbons in heavy oils. Future studies are expected to open a way to unite the dealkylation of alkyl polycyclic aromatic hydrocarbons with separation of aromatic compounds and alkanes, partial hydrogenation of polycyclic aromatics into tetralin derivatives, and ring-opening of tetralin derivatives into benzene derivatives, in order to maximize the yield of monocyclic aromatic compounds, i.e., benzene, toluene and xylene, and to maximize the possibility for use of alkyl parts as the diesel fuel.

Acknowledgement

A part of this work was carried out as a part of Research on Technologies for Advanced Use of Heavy Oil trusted by Ministry of Economy, Trade and Industry, Japan to Japan Petroleum Energy Center. The other part was supported by JSPS KAKENHI Grant Number 16H04568.

Table 1 Employed catalysts and their properties

Notation	Description	Composition	Number of Brønsted acid sites / mol kg^{-1}	Number of Lewis acid sites / mol kg^{-1}
MA	MoO_3 loaded on JRC-ALO-6*	Surface concentration of Si atom = 5 nm^{-2}	-	-
Alumina	JRC-ALO-6*	γ-Al_2O_3	0.02	0.22
SA*n*	Alumina-supported silica layer	Surface concentration of Si atom = *n* nm^{-2}	0.29 (SA11)	0.10 (SA11)
N631-L	Amorphous silica-alumina N631-L**	Al_2O_3 13 wt%, SiO_2 87 wt%	0.17	0.11
SAL-2	Amorphous silica-alumina JRC-SAL-2*	Al_2O_3 14 wt%, SiO_2 86 wt%	0.32	0.13
SAH-1	Amorphous silica-alumina JRC-SAH-1*	Al_2O_3 29 wt%, SiO_2 71 wt%	0.32	0.17
Y	In-situ H-form*** of Y zeolite, FAU structure	SiO_2/Al_2O_3 =5.9	-	-
N-USY	In-situ H-form*** of the steamed and NH_4NO_3-treated USY zeolite, FAU structure		0.51	0.20
β	H-form of Zeolite β, *BEA structure,	SiO_2/Al_2O_3 = 25	0.46	0.06

	JRC-Z-HB25 (1)*			
Mor-denite	H-form of mordenite, MOR structure, JRC-Z-HM20*	$SiO_2/Al_2O_3 = 20$	0.84	0.00
ZSM-5	In-situ H-form*** of ZSM-5 zeolite, MFI structure	$SiO_2/Al_2O_3 = 24$	1.15	0.00

* Reference catalyst supplied by Reference Catalyst Division, Catalysis Society of Japan.
** Kindly supplied by JGC Catalysts and Chemicals Ltd.
*** The NH_4-form was packed into the reactor and treated in a hydrogen flow at 773 K where the NH_4-form had been confirmed to be converted into the H-form.

Table 2 Catalytic activity for dealkylation of cumene

Catalyst	Reaction rate / mol_{cumeme} kg_{cat}^{-1}
Alumina	0.0003
SA12	0.055
N631-L	1.60
SAL-2	0.933
SAH-1	0.157

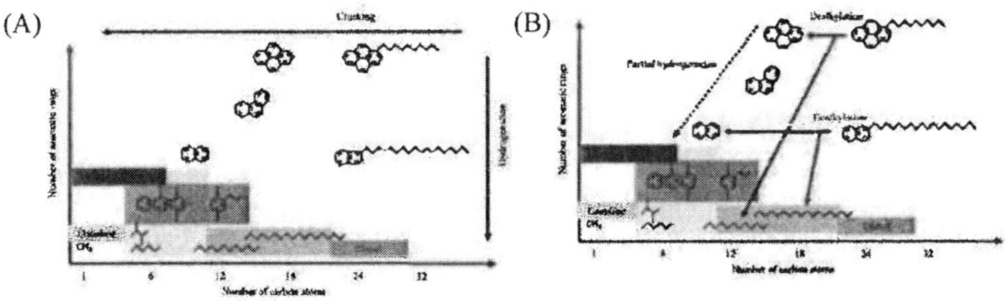

Figure 1 Schematic drawing of conventional hydrocracking (A) and dealkylation followed by partial hydrogenation of polyaromatics (B) of alkylaromatic hydrocarbons.

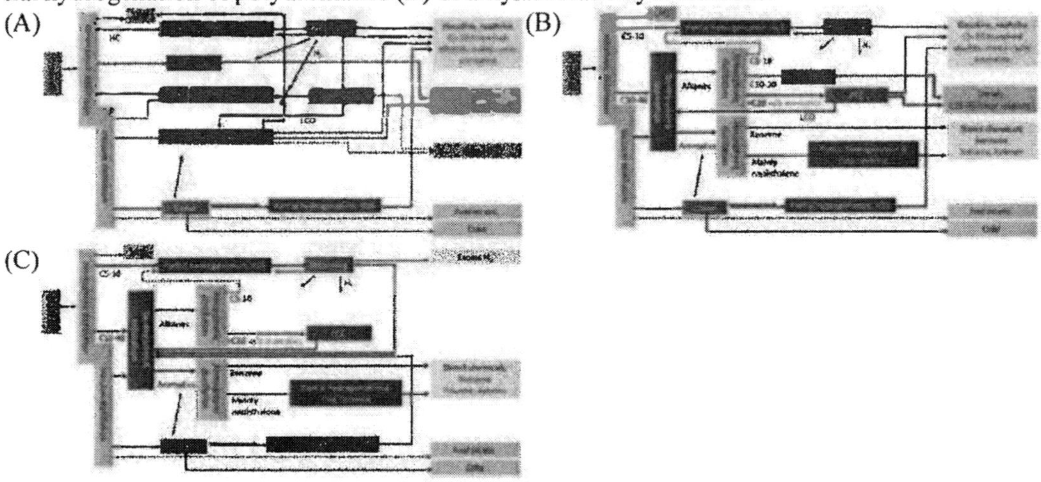

Figure 2 Schematic drawings of (A) conventional petroleum refinery, (B) what we aim with presently studied techniques in near future and (C) that in far future.

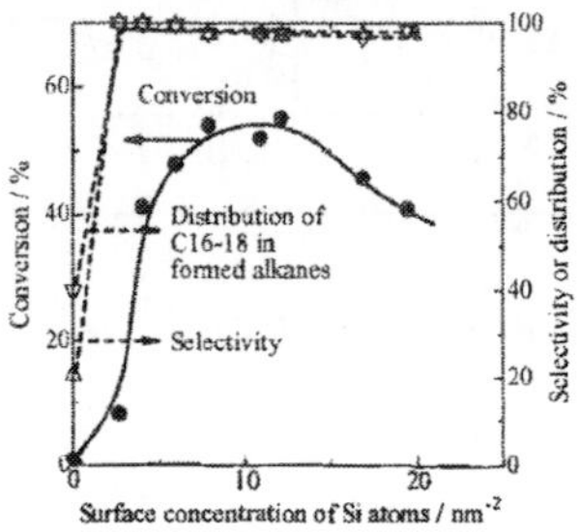

Figure 3 Plots of conversion (●), selectivity (△) and distribution of C16-18 alkanes in formed alkanes (▽) in dealkylation of C16, 18 AN reagent at 573 K and LHSV = 15 $g_{\text{-C16, 18 AN}}$ $g_{\text{-SA}}^{-1}$ h^{-1} in co-presence of MA (solid acid : MA weight ratio = 5 : 2) against amount of deposited silica on alumina.

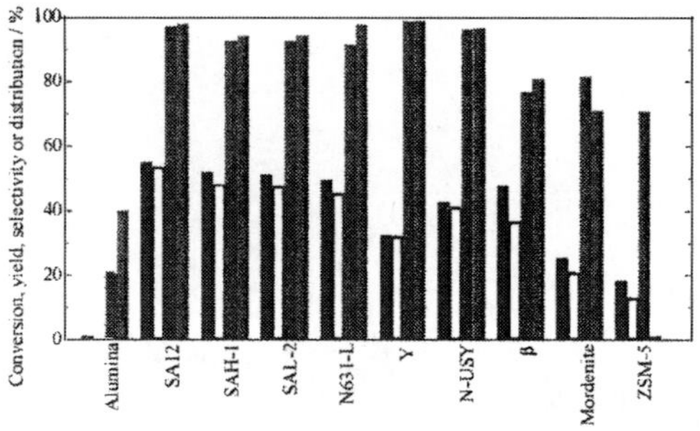

Figure 4 Conversion (▮), yield of desired products (▯), selectivity (▮) and distribution of C16-18 alkanes in formed alkanes (▮) in dealkylation of C16, 18 AN reagent at 573 K and LHSV = 15 $kg_{\text{C16, 18 AN}}$ kg_{cat}^{-1} h^{-1} in co-presence of MA (solid acid : MA weight ratio = 5 : 2) on various catalysts.

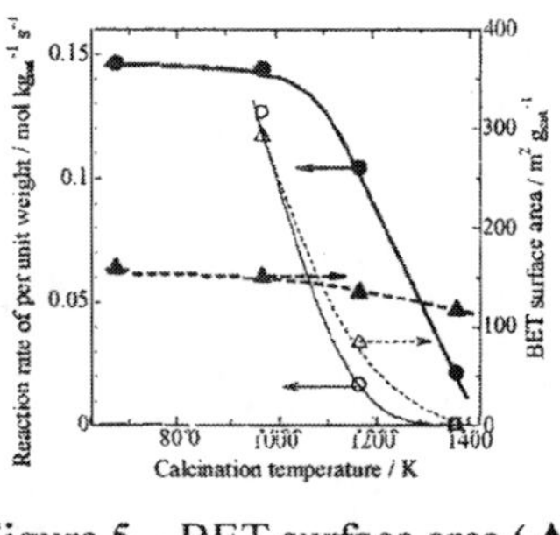

Figure 5 BET surface area (▲ and △) and reaction rate measured with adjusting catalyst amount to keep low conversion (< 17%) in dealkylation of C16 AN reagent (feed rate = 0.097 $g_{\text{-C16 AN}}$ h^{-1}) at 673 K (● and ○) on 15.5 wt% silica-deposited alumina (▲ and ●, SA10) and amorphous silica-alumina N631-L (△ and ○) calcined at various temperatures.

[1] R.A. Meyers, Handbook of Petroleum Refining Processes, Fourth Edition, McGraw-Hill Education, New York (2016).

[2] American Society of Testing Materials, Knocking Characteristics of Pure Hydrocarbons, ASTM Special Technical Publication No. 225 (1958).

[3] R. Kakizaki, M. Suzuki, T. Li, H. Ogawa, M. Murase, Trans. Soc. Automotive Engineers Jpn. 40 (2009) 1503-1508.

[4] M. Bender, Abstracts of DGMK International Conference, 2013, German Society for Petroleum and Coal, Science and Technology, Berlin (2013) 59-64.

[5] R. Sadeghbeigi, Fluid Catalytic Cracking Handbook, Third Edition, Elsevier, Amsterdam (2012).

[6] S.C. Korre, M.T. Klein, R.J. Quann, Ind. Eng. Chem. Res. 36 (1997) 2041-2050.

[7] W.-C. Cheng, G. Kim, A.W. Peters, X. Zhao, K. Rajagopalan, Catal. Rev.-Sci. Eng., 40 (1998) 39-79.

[8] C. Song, X. Ma, Appl. Catal., B: Environ. 41 (2003) 207-238.
[9] M.S. Rana, V. Sámano, J. Ancheyta, J.A.I. Diaz, Fuel 86 (2007) 1216-1231.
[10] J. Weitkamp, A. Raichle, Y. Traa, Appl. Catal., A: Gen. 222 (2001) 277-297.
[11] A. Raichle, Y. Traa, J. Weitkamp, Chem. Ing. Tech. 73 (2001) 947-956.
[12] X. Dupain, E.D. Gamas, R. Madon, C.P. Kelkar, M. Makkee, J.A. Moulijn, Fuel 82 (2003) 1559-1569.
[13] W.L. Orr, J.S.S. Damste, ACS Symp. Ser. 429 (1990) 2-29.
[14] G.S. Waldo, R.M.K. Carlson, J.M. Moldowan, K.E.Peters, J.E. Penner-Hahn, Geochim. Cosmochim. Acta, 55 (1991) 801-814.
[15] S. Mitra-Kirtley, O.C. Mullins, J.V. Elp., S.J. George, J. Chen., S.P. Cramer, J. Am. Chem. Soc. 115 (1993) 252-258.
[16] Q.Shi, D. Hou, K.H. Chung, C. Xu, S. Zhao, Y. Zhang, Energy & Fuels, 24 (2010) 2545-2553.
[17] M. Absi-Halabi, A. Stanislaus, D.L. Trimm, Appl. Catal. 72 (1991) 193-215.
[18] A.B. Amara, R. Dauphin, H. Babiker, Y. Viollet, J. Chang, N. Jeuland, A. Amer, Fuel 174 (2016) 63-75.
[19] P. Castaño, A.G. Gayubo, B. Pawelec, J.L.G. Fierro, J.M. Arandes, Chem. Eng. J. 140 (2008) 287-295.
[20] C. Song, Catal. Today, 86 (2005) 211-263.
[21] J. Weitkamp, ChemCatChem 4 (2012) 292-306.
[22] R. Pujro, M. Farco, U. Sedran, J. Chem. Technol. Biotechnol. 91 (2016) 336-345.
[23] X. Dupain, E.D. Gamas, R. Madon, C.P. Kelkar, M. Makkee, J.A. Moulijn, Fuel 82 (2003) 1559-1569.
[24] J. Ancheyta, S. Sánchez, M.A. Rodríguez, Catal. Today 109 (2005) 76-92.
[25] P. Castaño, A. Gutiérrez, I. Hita, J.M. Arandes, A.T. Aguayo, J. Bilbao, Energy & Fuels 26 (2012) 1509-1519.
[26] A. Gutiérrez, J.M. Arandes, P. Castaño, M. Olazar, A. Barona, J. Bilbao, Fuel Proc. Technol. 95 (2012) 8-15.
[27] A. Gutiérrez, J.M. Arandes, P. Castaño, A.T. Aguayo, J. Bilbao, Energy & Fuels 25 (2011) 3389-3399.
[28] A. Gutiérrez, J.M. Arandes, P. Castaño, M. Olazar, J. Bilbao, Fuel Proc. Technol., 101, 2012, 64-72 (2012).
[29] I. Gawel, D. Bociarska, P. Biskupski, Appl. Catal., A: Gen. 295 (2005) 89-94.
[30] T. Nakanishi, R. Komiyama, Supply and Demand Analysis on Pertoleum Products and Crude Oils for Asia and the World, Inst. Energy Economy Jpn. (2006).
[31] B. Sirjean, P.A. Glaude, M.F. Ruiz-Lopez, R. Fournet, J. Phys. Chem., A 110 (2006) 12693-12704.
[32] A. Corma, B.W. Wojciechowski, Catal. Rev. Sci. Eng. 24 (1982) 1-65.
[33] A.M. Youssef, A.I. Ahmed, S.E. Samra, Mater. Lett. 10 (1990) 175-180.
[34] G.W. Cassell, N. Dural, A.L. Hines, Ind. Eng. Chem. Res. 28 (1989) 1369-1374.
[35] S.J. Ardakani, K.J. Smith, Appl. Catal. A: Gen. 403 (2011) 36-47.
[36] T. Choi, J. Lee, J. Shin, S. Lee, D. Kim, J.K. Lee, Appl. Catal. A: Gen. 492 (2015) 140-150.
[37] P. Castaño, J.M. Arandes, M. Olazar, J. Bilbao, B. Pawelec, U. Sedrán, Catal. Today 150 (2010) 363-367.
[38] R. Contreras, J. Ramírez, R. Cuevas-Carcía, A. Gutiérrez-Alejandre, P. Castillo-Villalón, G. Macías, I. Puente-Lee, Catal. Today 148 (2009) 49-54.
[39] J.L. Hodala, A.B. Halgeri, G.V. Shanbhag, Appl. Catal. A: Gen. 484 (2014) 8-16.
[40] H. Liu, G. Yang., React. Kinet. Mech. Catal. 113 (2014) 605-614.
[41] S.A. Ali, A.M. Aitani, J. Čejka, S.S. Al-Khattaf, Catal. Today 243 (2015) 118-127.
[42] C. Hu, J. Li, W. Jia, M. Liu, Z. Hao, Z. Zhu Ghin. J. Chem. 33 (2015) 247-252.
[43] A.L. Petre, W.F. Hoelderich, M.L. Gorbaty, Appl. Catal. A: Gen. 363 (2009) 100-108.
[44] R. Brzowski, W. Skupiński, J. Catal. 220 (2003) 13-22.
[45] C.M. Smith, P.E. Savage, Ind. Eng. Chem. Res. 30 (1991) 331-339.

---

[46] J. Beens, U.A.T. Brinkman, Anal. Biochem. Chem. 378 (2004) 1939-1943.
[47] T. Górecki, J. Harynuk, O. Panić, J. Sep. Sci. 27 (2004) 359-379.
[48] J.H. Marsman, J. Wildschut, F. Mahfud, H.J. Heeres, J. Chromatogr., A 1150 (2007) 21-27.
[49] T. Jin, T. Okuhara, J.M. White, J. Chem. Soc., Chem. Commun. (1987) 1248-1249.
[50] S. Sato, M. Toita, Y.-Q. Yu, T. Sodesawa, F. Nozaki, Chem. Lett. (1987) 1535-1536.
[51] M. Niwa, T. Hibino, H. Murata, N. Katada, Y. Murakami, J. Chem. Soc., Chem. Commun. (1989) 289 - 290.
[52] B.-Q. Xu, T. Yamaguchi, K. Tanabe, Chem. Lett. (1989) 149-152.
[53] H. Fukui, T. Ogawa, M. Nakano, M. Yamaguchi, Y. Kanda, in Controlled Interphases in Composite Materials, ed. by H. Ishida, Springer, Berlin (1990) 469-478.
[54] B. Beguin, E. Grabowski, M. Primet, J. Catal. 127 (1991) 595-604.
[55] N. Okazaki, T. Kohno, R. Inoue, Y. Imizu, A. Tada, Chem. Lett. (1993) 1195-1198.
[56] P. Sarrazin, S. Kasztelan, N. Zanier-Szyndlowski, J.P. Bonnelle, J. Grimblot, J. Phys. Chem. 97 (1993) 5947-5953.
[57] T. Horiuchi, T. Sugiyama, T. Mori, J. Mater. Chem. 3 (1993) 861-865.
[58] T.-C. Sheng, I.D. Gay, J. Catal. 145 (1994) 10-15.
[59] M. Niwa, N. Katada, Y. Murakami, J. Phys. Chem. 94 (1990) 6441-6445.
[60] N. Katada, H. Ishiguro, K. Muto, M. Niwa, Adv. Mater. Chem. Vap. Deposition 1 (1995) 54-60.
[61] N. Katada, T. Tsubouchi, T. Toyama, M. Niwa, Y. Murakami, Res. Chem. Intermed., 9 (1995) 137-149.
[62] N. Katada, T. Toyama, M. Niwa, J. Phys. Chem., 98 (1994) 7647-7652.
[63] M. Niwa, K. Suzuki, N. Katada, T. Kanougi, T. Atoguchi, J. Phys. Chem. B 109 (2005) 18749-18757.
[64] M. Niwa, N. Katada, K. Okumura, Characterization and Design of Zeolite Catalysts: Solid Acidity, Shape Selectivity and Loading Properties, Springer, Berlin, Heidelberg, Dordrecht and New York (2010).
[65] K. Okumura, T. Tomiyama, N. Morishita, T. Sanada, K. Kamiguchi, N. Katada, M. Niwa, Appl. Catal. A: Gen. 405 (2011) 8-17.
[66] S. Brunauer, P.H. Emmett, E. Teller, J. Am. Chem. Soc. 60 (1938) 309-319.
[67] E.P. Barrett, L.G. Joyner, P.P. Halenda, J. Am. Chem. Soc. 73 (1951) 373-380.
[68] A. Galarneau, F. Villemot, J. Rodriguez, F. Fajula, B. Coasne, Langmuir 30 (2014) 13266-13274.
[69] M. Niwa, S. Inagaki, Y. Murakami, J. Phys. Chem. 89 (1985) 3869-3872.
[70] N. Katada, T. Tsubaki, M. Niwa, Appl. Catal. A: Gen. 340 (2008) 76-86.
[71] D.W. Breck, Zeolite Molecular Sieves: Structure, Chemistry and Use, John Wiley and Sons, New York (1974).
[72] N. Katada, M. Niwa, Catal. Surveys Asia 8 (2004) 161-170.
[73] N. Katada, K. Suzuki, T. Noda, G. Sastre, M. Niwa, J. Phys. Chem. C 113 (2008) 19208-19217.
[74] N. Katada, T. Fujii, K. Iwata, Y. Hibino, M. Niwa, J. Catal. 186 (1999) 478-480.
[75] K. Muto, N. Katada, M. Niwa, Catal. Today 35 (1997) 145-151.
[76] N. Katada, Y. Kawaguchi, K. Takeda, T. Matsuoka, N. Uozumi, K. Kanai, S. Fujiwara, K. Kinugasa, K. Nakamura, S. Suganuma, M. Nanjo, Appl. Catal. A: Gen. 530 (2017) 93-101.
[77] K. Kinugasa, F. Nakano, S. Nagano, S. Suganuma, E. Tsuji, N. Katada, J. Jpn. Petrol. Inst. 61 (2018) 294-301.

Abstract title: Selective Aerobic Oxidation of Hydrocarbons by N-Hydroxyphthalimide-Based Catalyst for Cetane Enhancement

Authors: Maryam Altaher, Christos Kalamaras, Mohammed Alqahtani

Affiliation: Saudi Aramco

Abstract

Gasoline compression ignition (GCI) utilizes gasoline-like fuels in low temperature compression ignition (CI) engines. As a result, thermal efficiency is increased comparing to conventional spark ignition (SI) engine, while NOx and soot tailpipe emissions are reduced comparing to diesel engine. However, this combustion strategy requires gasoline fuel with moderate ignition delay for cold start and high load demand conditions. This work investigates a process to control the ignition characteristics of gasoline to widen the GCI engine operation conditions. Selective aerobic oxidation of hydrocarbons using N-hydroxyphthalimide (NHPI) based catalyst under moderate reaction conditions has been suggested as an effective method to produce high cetane components like peroxides.

Toluene was used as a representative molecule for this work, since aromatics were the only hydrocarbon family that found to produce peroxides under the studied conditions. The produced peroxide concentration was quantified by titration using triphenylphosphine (Ph3P) followed by Gas Chromatography (GC) analysis. A detailed parametric study was performed using autoclave batch reactor to optimize the reaction conditions as a function of peroxide yields. In addition, supported NHPI catalysts over several porous materials were synthesized and screened. Thermal Gravimetric Analysis (TGA) was used to quantify the amount of NHPI loaded on the supports. It's been found that bare NHPI enhanced the formation of peroxide twice than supported NHPI. The bare NHPI was proved to be a robust catalyst that enhanced the cetane number by 30%. Selective aerobic oxidation of toluene using NHPI-based catalysts is found to be a promising technology to control the ignition characteristics of fuel and hence enables GCI engine concept.

Catalytic technology for production of C4 chemicals via erythritol from biomass

Keiichi Tomishige
Department of Applied Chemistry, School of Engineering, Tohoku University
Aoba 6-6-07, Aramaki, Aoba-ku, Sendai 980-8579, Japan
E mail: tomi@erec.che.tohoku.ac.jp

Abstract

C4 chemicals are one of building blocks in the petrochemical industry for production of polymers and so on. At present, C4 chemicals have been derived from naphtha, which will be decreased by the shift to gas industries. Therefore, the synthesis of C4 chemicals from biomass can be a promising method for the substitution of naphtha. It has been known that erythritol is produced by the fermentation of glucose and glycerol. At the same time, the dehydration of erythritol can give 1,4-anhydroerythritol in high yield. These suggest that erythritol and 1,4-anhydroerythritol can be utilized as biomass-derived C4 building blocks. Our group has been developed heterogeneous catalysts for the selective conversion of erythritol and 1,4-anhydroerythritol using H_2 as a reductant to various C4 chemicals such as 1,4-, 1,2-, and 1,3-butanediols, butanols, butadiene and so on, where C-O hydrogenolysis, deoxydehydration, etc. are catalyzed.

1. Introduction

C4 chemicals have been produced from petroleum-based raw materials such as butadiene, butane, ethylene and so on. Butanediols are regarded as one of the value-added chemicals. A practical method for the production of 1,4-butanediol from 1,3-butadiene is shown below.

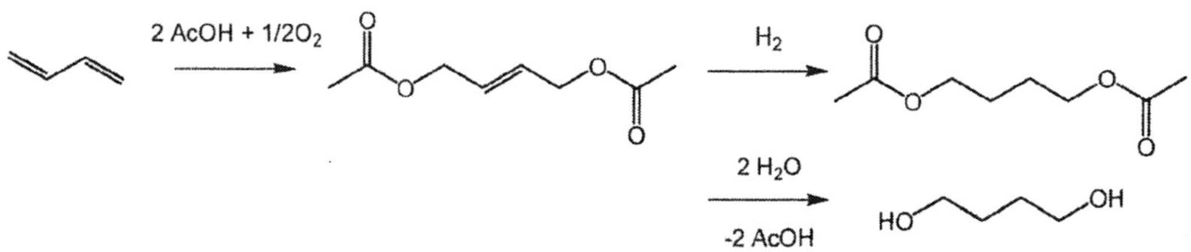

Another method for the production of 1,4-butanediol via maleic anhydride has been also utilized as shown below.

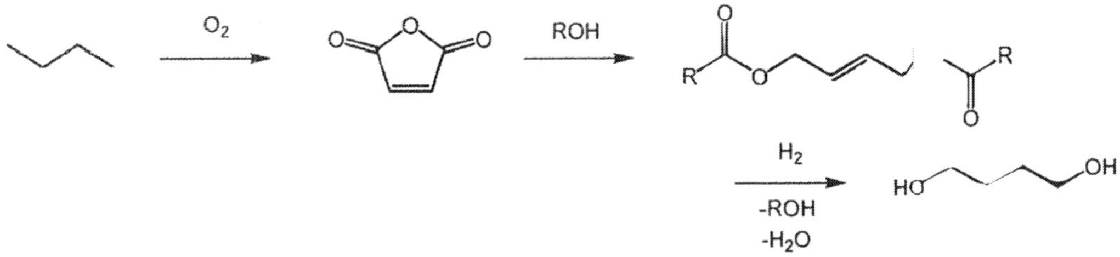

Various kinds of chemicals have been produced via 1,4-butanediol and examples are shown below.

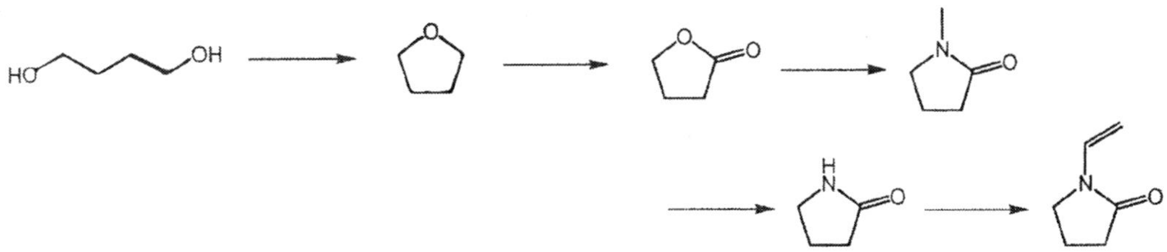

On the other hand, 1,3-butanediol has been produced from acetaldehyde as shown below.

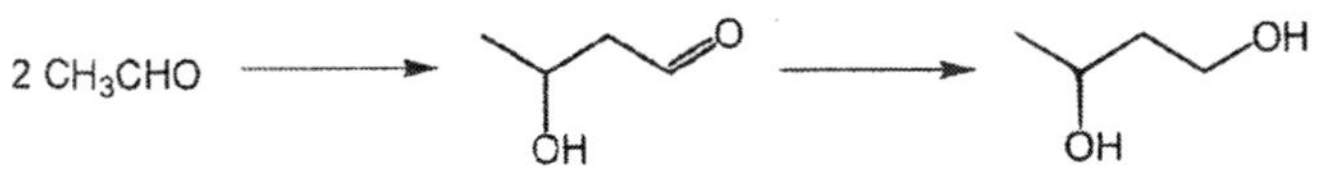

As shown above, petroleum-based raw materials have low oxygen content, the oxygen-containing chemicals such as butanediols are synthesized by the addition of oxygen atoms to hydrocarbons, for example, oxidation with air. On the other hand, the building blocks in the biomass refinery have very high oxygen content as shown below.

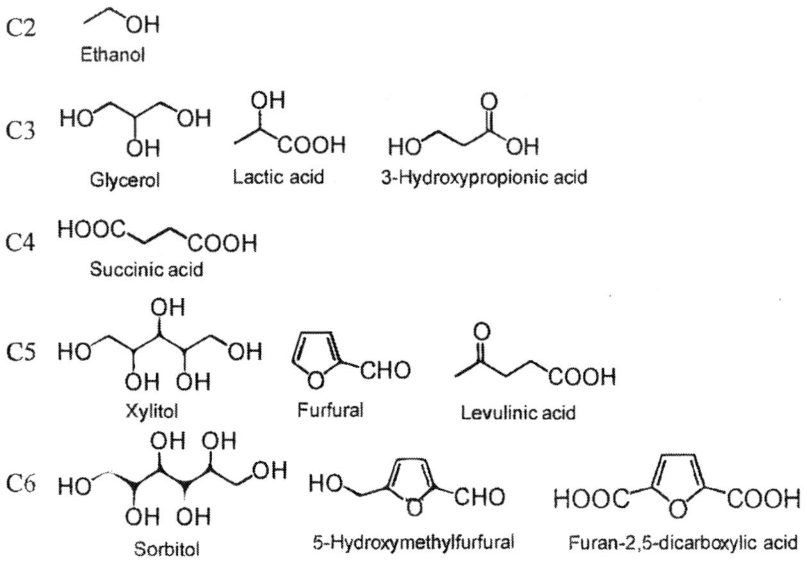

These building blocks can be derived relatively easily by the fermentation and/or chemical transformation from biomass such as cellulose, hemicellulose, vegetable oils and so on. The C4 building block is succinic acid, which can be produced by the fermentation of glucose. Hydrogenation of succinic acid can give γ-butyrolactone and 1,4-butanediol. Generally, carboxylic acids are converted to corresponding alcohols via esters as shown above. In order to enhance the feasibility of the hydrogenation of carboxylic acids, the catalysts have been development for the direct hydrogenation of carboxylic acids without esterification [1-3]. However, C4 building block is so limited and it is not easy to synthesis C4 compounds with functional groups at 2- or 3-positions from succinic acid. Therefore, another candidate is expected to be "erythritol" as a C4 building block because erythritol is reported to be produced by the fermentation of glucose and glycerol and all the carbon atoms are functionalized [4]. At the same time, erythritol is also converted to 1,4-anhydroerythritol selectively. Here, I would like to introduce the catalytic technologies for the conversion of erythritol and 1,4-anhydroerythritol to various value-added chemicals as shown in Figure 1.

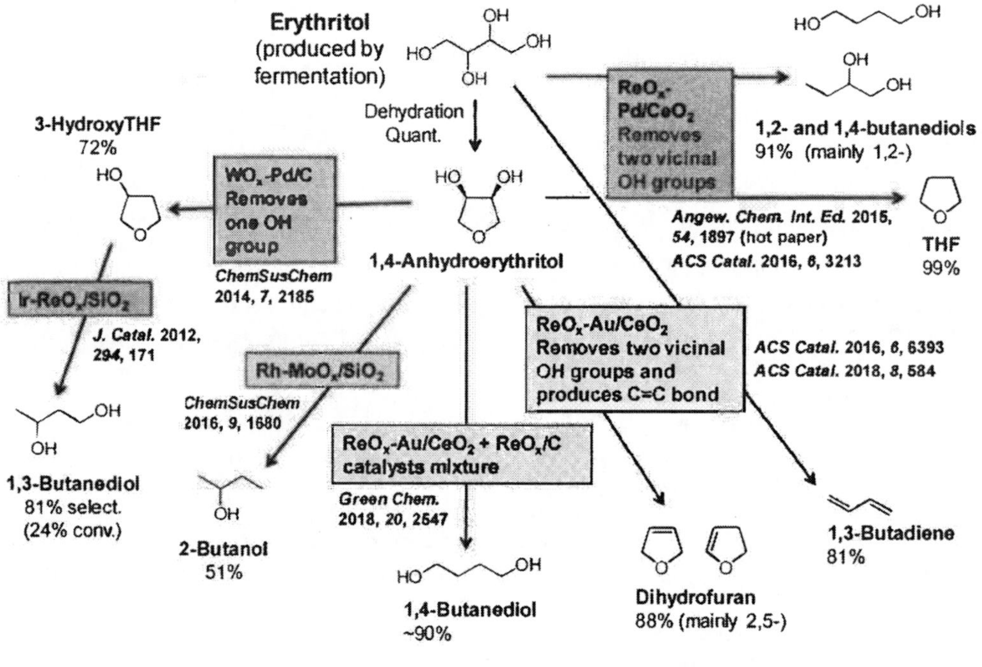

Figure 1. Selective productions of C4 chemicals from erythritol.

2. C-O hydrogenolysis of erythritol, 1,4-anhydroerythritol, and their derivatives

Typical reactions of C-O hydrogenolysis is divided into two categories: one is the hydrogenolysis of C-OH with H_2 to give C-H and H_2O, and the other is hydrogenolysis of C-O-C with H_2 to give C-H and C-OH. The hydrogenolysis of C-OH is also regarded as hydrodeoxygenation. Generally, the oxygen content of biomass-derived substrates is clearly higher than value-added products which have been produced from petroleum, and the C-O hydrogenolysis and hydrodeoxygenation is more important [5-13].

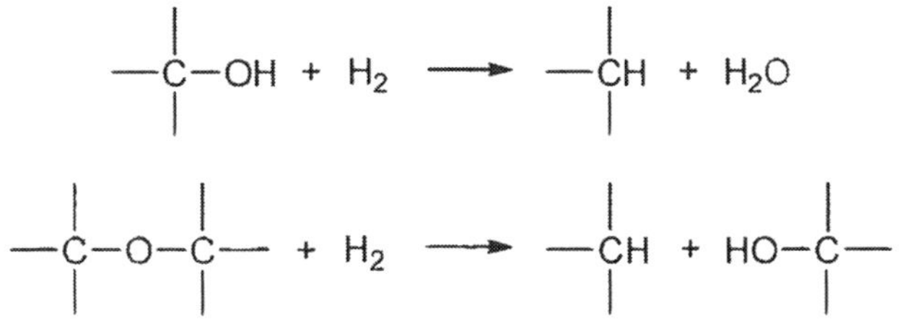

Bimetallic Ir-ReO$_x$/SiO$_2$ and Rh-MoO$_x$/SiO$_2$ catalysts showed high catalytic performance in the C-O hydrogenolysis of various biomass-derived substrates such as glycerol, tetrahydrofurfuryl alcohol and so on [5-13]. Ir-ReO$_x$ and Rh-MoO$_x$ based catalysts showed high performance in the hydrogenolysis of glycerol and tetrahydrofurfuryl alcohol [14-18], Based on these results, Ir-ReO$_x$/SiO$_2$ catalyst was applied to the hydrogenolysis of erythritol, where the target products were butanediols. A typical result is shown below [19].

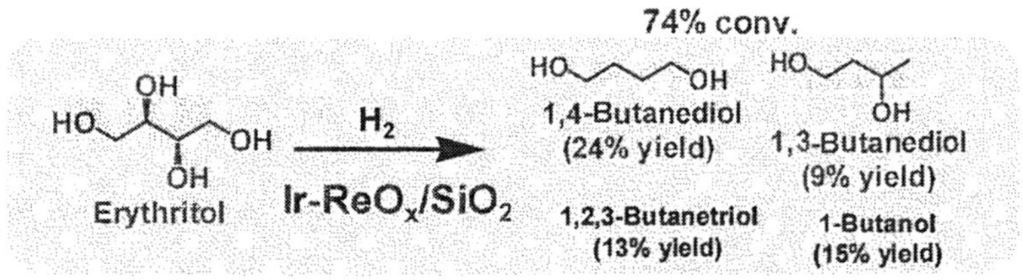

The yield of 1,4- and 1,3-butanediol is not high enough. In particular, the two-time C-O hydrogenolysis is necessary in the conversion of erythritol to butanediols, however, high selectivity and yield of a specific target product such as 1,4-butanediol was not achieved on Ir-ReO$_x$/SiO$_2$. Another approach is the conversion of erythritol and 1,4-anhydroerythritol by C-O hydrogenolysis to 2-butanol. This is based on the very low reactivity of secondary monoalcohol over Ir-ReO$_x$, Rh-ReO$_x$, and Rh-MoO$_x$ catalysts [20-23], which can be connected to high yield of secondary monoalcohol as a target product. The highest yield of 2-butanol from the hydrogenolysis of erythritol using Rh-MoO$_x$/SiO$_2$ was 34%, and that of 2-butanol was improved to be 51% by using 1,4-anhydroerythritol [24]. Regarding the conversion of 1,4-anhydroerthritol, we also found that WO$_x$-Pd/C gave 72% yield of 3-hydroxytetrahydrofuran from 1,4-anhydroerythritol [25]. On the other hand, the hydrogenolysis of 3-hydroxytetrahydrofuran using Ir-ReO$_x$/SiO$_2$ was investigated and it is found that 1,3-butanediol was given in high selectivity [26], where 81% selectivity at 24%

conversion. These results suggest the two-step conversion of 1,4-anhydroerythritol to 1,3-propanediol via 3-hydroxytetrahydrofuran.

3. Deoxydehydration of erythritol, 1,4-anhydroerythritol, and their derivatives

Deoxydehydration is the removal of vicinal OH groups to give carbon-carbon double bond as described below.

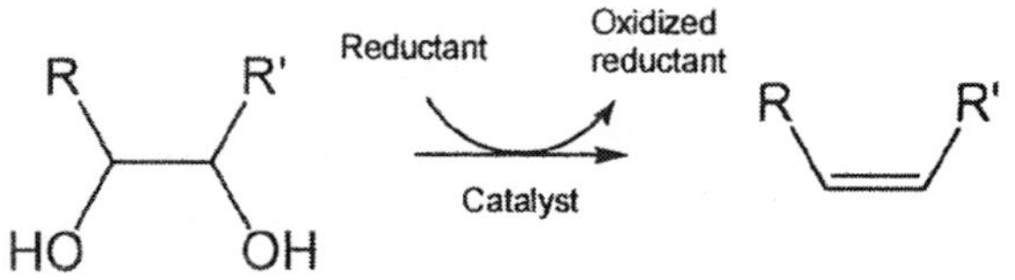

It has been reported that homogeneous Re(VII) complex such as $C_5Me_5ReO_3$ and CH_3ReO_3 catalyzes the deoxydehydration, however these catalyst systems demanded PPh_3 and secondary alcohols as a reductant because these Re complexes have low ability to activate H_2, although H_2 is one of the cheapest reductant. Recently our groups have reported the heterogeneous Re catalysts for the deoxydehydration such as ReO_x-Pd/CeO_2 and ReO_x-Au/CeO_2 [27-30]. One of the important characters is that there two catalysts enabled the utilization of H_2 as a reductant. Hydrogen molecules can be activated on the metal surfaces of Pd or Au and the dissociated hydrogen species can move to Re species and to reduce Re species attached on CeO_2 surface. ReO_x-Pd/CeO_2 catalyzed the deoxydehydration of vicinal OH groups and the consecutive hydrogenation of C=C in the deoxydehydration product. In contrast, the surface of Au can activate hydrogen molecular, however, the activity of C=C hydrogenation is rather low, and ReO_x-Au/CeO_2 catalyzed the deoxydehydration without the consecutive hydrogenation of C=C in the deoxydehydration product.

Deoxydehydration of erythritol with H_2 catalyzed by ReO_x-Pd/CeO_2 gave 1,2- and 1,4-butanediols. In particular, the yield of 1,2- and 1,4-butanediols is 77% and 12%, respectively, indicating that the deoxydehydration of vicinal primary-secondary OH groups is more favorable than that of vicinal secondary-secondary OH groups. As mentioned above, the C-O hydrogenolysis of erythritol to the specific butanediol is not easy and the yield of 1,4-butanediol is not high. In contrast, deoxydehydration+hydrogenation catalyzed by ReO_x-Pd/CeO_2 gave 1,2-butanediol in high yield (77%).

Figure 2 shows the reaction mechanism of deoxydehydration+hydrogenation over ReO_x-Pd/CeO_2 [28]. The catalytic cycle consists of the redox one between Re(VI) and Re(IV).

One of the roles of Pd is to promote the reduction of Re(VI) to Re(IV) with the activation of H_2. The role of CeO_2 is to stabilize the Re(IV) species and suppress the overreduction of Re species which gives inactive species for deoxydehydration. The substrate interacts with Re species give a diolate, and the reaction of the diolate to Re(VI) and the deoxydehydration product is a rate-determining step on the basis of the kinetic analysis. The deoxydehydration product is hydrogenated over Pd to give corresponding saturated product.

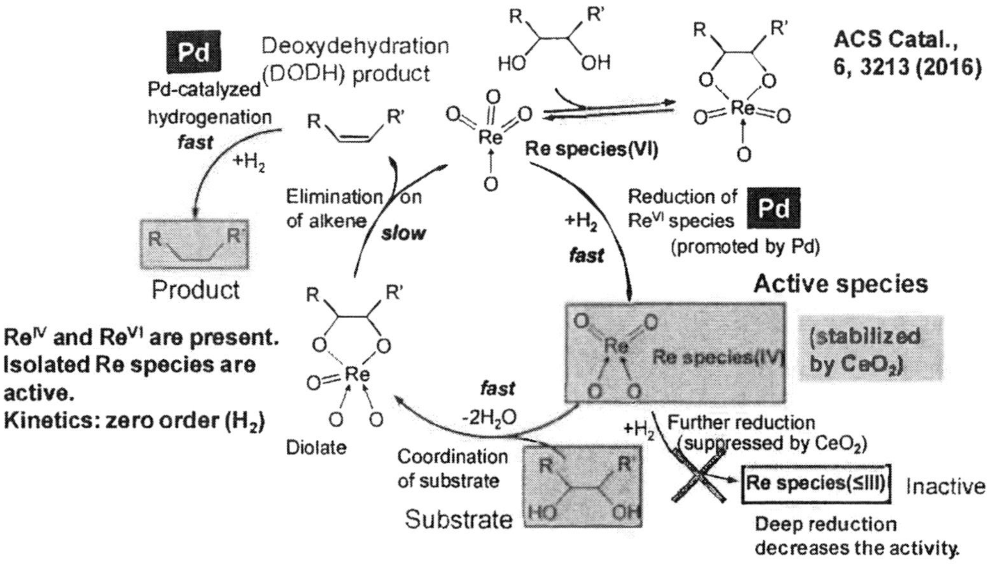

Figure 2. Catalytic cycle of deoxydehydration+hydrogenation of the substrate with vicinal diols with H_2 over ReO_x-Pd/CeO_2.

Deoxydehydration+hydrogenation of 1,4-anhydroerythritol with H_2 catalyzed by ReO_x-Pd/CeO_2 gave tetrahydrofuran in very high yield (>99%) [27]. The turnover number of tetrahydrofuran formation on the basis of Re amount over ReO_x-Pd/CeO_2 in the reaction of 1,4-anhydroerythritol with H_2 was clearly higher than that over homogeneous CH_3ReO_3 in the reaction with 2-pentanol at the same reaction temperature [31], indicating the very high activity of ReO_x-Pd/CeO_2 as a heterogeneous deoxydehydration catalyst. Very high catalytic activity of ReO_x/CeO_2 modified with Pd or Au can decrease the reaction temperature and enable the deoxydehydration of methyl glycosides with cis-vicinal OH groups to corresponding dideoxysugars and unsaturated sugars [32, 33].

Deoxydehydration of erythritol with H_2 catalyzed by ReO_x-Au/CeO_2 gave 1,3-butadiene in high yield (81%) [29]. Deoxydehydration of 1,4-anhydroerythritol with H_2 catalyzed by ReO_x-Au/CeO_2 gave dihydrofurans (2,5-dihydrofuran as a main product, and 2,3-dihydrofuran) in high yield (88%) [29], which is an important result from the viewpoint of further transformation of 2,5-dihydrofuran. Figure 3 shows model structure of ReO_x-$^{imp}Au^{0.3}$/CeO_2 (Au/Re=0.3, Re 1 wt%, Au: impregnation method) [30]. It is clear that the number of Au particles is much smaller than that of CeO_2 where Re species are highly dispersed. However, very small number of Au particles actually promoted the deoxydehydration over most Re species, probably by hydrogen spillover phenomenon from Au to CeO_2 and Re. The mechanism of the Re reduction by hydrogen spillover is also illustrated in Figure 3.

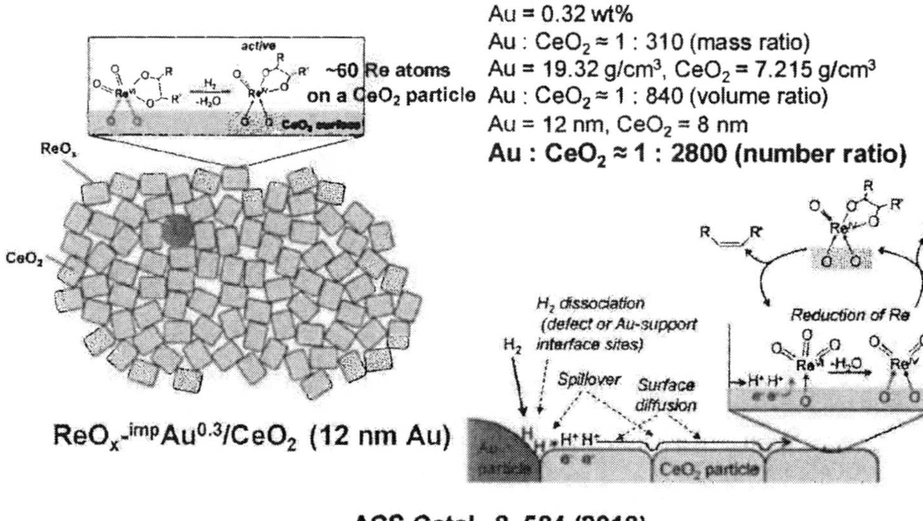

Figure 3. Model Structure of ReO_x-$^{imp}Au^{0.3}$/CeO_2 (Au/Re=0.3, Re 1 wt%, Au: impregnation method) and mechanism of the Re reduction of by small number of Au particles

It has been reported that ReO_x/TiO_2 catalyzed the conversion of 2,5-dihydrofuran to 1,4-butanediol [34]. Inspired by this report, we develop the one-pot conversion of 1,4-anhydroerythritol to 1,4-butanediol using the combination of ReO_x-Au/CeO_2 with other Re catalyst which has the ability to convert 2,5-dihydrofuran to 1,4-butanediol. We investigated the support effects of Re catalysts for the cocatalyst in the combination with

ReO_x-Au/CeO_2. As a result, ReO_x/C catalysts were found to be effective as a cocatalyst in the one-pot conversion of 1,4-anhydroerythritol to 1,4-butanediol, and the yield of 1,4-butanediol was as high as 90% [35]. In order to elucidate the role of ReO_x/C cocatalyst, the reactivity and selectivity of products of the intermediates such as 2,5- and 2,3-dihydrofurans over ReO_x/C, C and so on. The reaction scheme of one-pot conversion of 1,4-anhydroerythritol to 1,4-butanediol over ReO_x-Au/CeO_2+ReO_x/C physical mixture catalyst is shown in Figure 4.

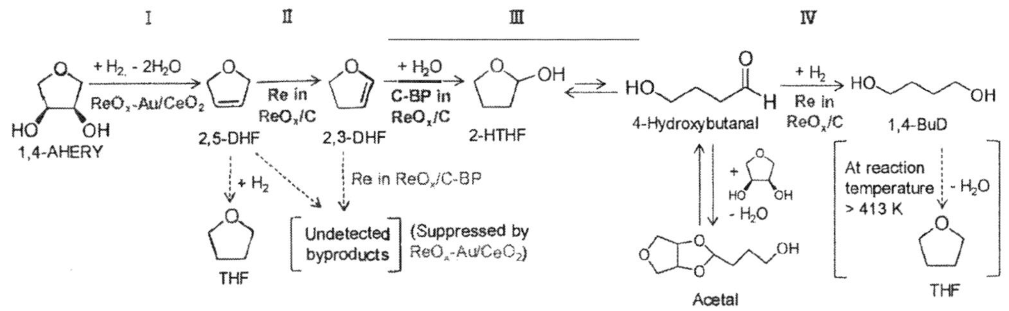

AHERY: anhydroerythritol, BuD: butanediol, THF: tetrahydrofuran, DHF: dihydrofuran, GBL: γ-butyrolactone, BuOH: butanol, 2-HTHF: 2-Hydroxy-tetrahydrofuran.

Figure 4. Reaction scheme of one-pot conversion of 1,4-anhydroerythritol to 1,4-butanediol over ReO_x-Au/CeO_2+ReO_x/C physical mixture catalyst

As mentioned above, 1,4-anhydroerythritol is converted to 2,5-dihydrofuran over ReO_x-Au/CeO_2, and Re species on ReO_x/C catalyzed the isomerization of 2,5-dihydrofuran to 2,3-dihydrofuran. Hydration of 2,3-dihydrofuran to 2-hydroxytetrahydrofuran is catalyzed by the surface of carbon-support. An important point is that 1,4-dioxane solvent was used for this one-pot conversion and H_2O was derived only from the deoxydehydration of 1,4-anhydroerythritol and the concentration of H_2O is very low during the one-pot reaction. Therefore, carbon support can catch the low concentration of H_2O and utilize this for the hydration of 2,3-dihydrofuran. 2-Hydroxytetrahydrofuran has a ring-opening structure (4-hydroxybutanl) at the equilibrium, and 4-hydroxybutanal is hydrogenated to 1,4-butanediol catalyzed by ReO_x/C. Considering that hydrogenation of carbonyl groups is catalyzed by metal surface, a part of Re species on carbon support is reduced to the metallic state. In the case of ReO_x/SiO_2, it has been reported that Re species on SiO_2 was reduced to the metallic state with gas phase H_2 and in the hydrogenation of carboxylic acids using 1,4-dioxane solvent, where the reaction conditions are similar to those in the one-pot conversion of 1,4-anhydroerythritol [2, 36]. Totally, ReO_x/C was regarded as a multifunctional catalyst in the one-pot conversion of 1,4-anhydroerythritol by the combination with ReO_x-Au/CeO_2. In addition, we found that 1,4-butanediol can be produced from 1,4-anhydroerythritol through the co-catalysis of monometallic mixed catalysts ($ReO_x/CeO_2+ReO_x/C$) in the one-pot reduction with H_2 [37]. The highest yield of 1,4-butanediol was over 80%, which is similar to the value obtained over ReO_x-$Au/CeO_2+ReO_x/C$ catalysts. Mixed catalysts of CeO_2+ReO_x/C showed almost the same performance, giving 89% yield of 1,4-butanediol. These indicate that high valent Re species are mobile, in particular, before the reduction. The reduction of Re species can proceed during the heating in the presence of high-pressure hydrogen. A part of Re species on the carbon support is reduced to the metallic state, and this metallic Re species also play a crucial role in the activation of H_2 which is the role of Au in the case of ReO_x-$Au/CeO_2+ReO_x/C$. This means that it is possible to design the catalysts without using Au.

4. Conclusions

Important C4 chemicals such as 1,4-, 1,3-, and 1,2-butanediols, 1,3-butadiene, tetrahydrofuran, dihydrofuran, 2-butanol and so on can be derived from erythritol and 1,4-anhydroerythritol using suitable catalysts and H_2 as a reductant. Erythritol can be synthesized from glycerol by fermentation, so it can be regarded as a building block in the biomass refinery. At present, the examples of erythritol conversion are so limited, and further investigation on the catalyst development for the selective conversion of erythritol will contribute to the sustainable society.

Acknowledgements

A part of this work is supported by JSPS KAKENHI Grant Number 18H05247.

References

1. Y. Takeda, Y. Nakagawa, K. Tomishige, Catal. Sci. Technol., 2, 2221 (2012).
2. Y. Takeda, M. Tamura, Y. Nakagawa, K. Okumura, K. Tomishige, ACS Catal., 5, 7034 (2015).
3. Y. Takeda, M. Tamura, Y. Nakagawa, K. Okumura, K. Tomishige, Catal. Sci. Technol., 6, 5668 (2016).
4. Y. Kobayashi, H. Iwata, D. Mizushima, J. Ogihara, T. Kasumi, Lett. Appl. Microbiol., 60, 475 (2015).
5. Y. Nakagawa, M. Tamura, K. Tomishige, Fuel Proc. Technol., 193, 404 (2019).
6. Y. Nakagawa, M. Tamura, K. Tomishige, Res. Chem. Intermed., 44, 3879 (2018).
7. K. Tomishige, Y. Nakagawa, M. Tamura, Green Chem., 19, 2876 (2017).
8. Y. Nakagawa, M. Tamura, K. Tomishige, Catal. Surv. Asia, 19, 249 (2015).
9. Y. Nakagawa, M. Tamura, K. Tomishige, ChemSusChem, 8, 1114–1132 (2015).
10. K. Tomishige, M. Tamura, Y. Nakagawa, Chemical Record, 14, 1041 (2014).

11. Y. Nakagawa, M. Tamura, K. Tomishige, J. Mater. Chem. A, 2, 6688 (2014).
12. Y. Nakagawa, K. Tomishige, Catal. Today, 195, 136 (2012).
13. Y. Nakagawa, K. Tomishige, Catal. Surv. Asia, 15, 111 (2011).
14. L. Liu, S. Kawakami, Y. Nakagawa, M. Tamura, K. Tomishige, Appl. Catal. B, 256, 117775 (2019).
15. Y. Amada, H. Watanabe, M. Tamura, Y. Nakagawa, K. Okumura, K. Tomishige, J. Phys. Chem. C, 116, 23503 (2012).
16. S. Liu, M. Tamura, Z. Shen, Y. Zhang, Y. Nakagawa, K. Tomishige, Catal. Today, 303, 106 (2018)
17. S. Liu, Y. Amada, M. Tamura, Y. Nakagawa, K. Tomishige, Catal. Sci. Technol., 4, 2535 (2014).
18. S. Koso, Y. Nakagawa, K. Tomishige, J. Catal., 280, 221 (2011).
19. Y. Amada, H. Watanabe, Y. Hirai, Y. Kajikawa, Y. Nakagawa, K. Tomishige, ChemSusChem, 5, 1991 (2012).
20. K. Chen, M. Tamura, Z. Yuan, Y. Nakagawa, K. Tomishige, ChemSusChem, 6, 613-621 (2013).
21. S. Liu, M. Tamura, Y. Nakagawa, K. Tomishige, ACS Sustainable Chem. Eng., 2, 1819 (2014).
22. S. Liu, Y. Okuyama, M. Tamura, Y. Nakagawa, A. Imai, K. Tomishige, ChemSusChem, 8, 628 (2015).
23. S. Liu, Y. Okuyama, M. Tamura, Y. Nakagawa, A. Imai, K. Tomishige, Green Chem., 18, 165 (2016).
24. T. Arai, M. Tamura, Y. Nakagawa, K. Tomishige, ChemSusChem, 9, 1680 (2016).
25. Y. Amada, N. Ota, M. Tamura, Y. Nakagawa, K. Tomishige, ChemSusChem, 7, 2185 (2014).
26. K. Chen, K. Mori, H. Watanabe, Y. Nakagawa, K. Tomishige, J. Catal., 243, 171 (2012).
27. N. Ota, M. Tamura, Y. Nakagawa, K. Okumura, K. Tomishige, Angew. Chem. Int. Ed., 54, 1897 (2015).
28. N. Ota, M. Tamura, Y. Nakagawa, K. Okumura, K. Tomishige, ACS Catal., 6, 3213 (2016).
29. S. Tazawa, N. Ota, M. Tamura, Y. Nakagawa, K. Okumura, K. Tomishige, ACS Catal., 6, 6393 (2016).
30. Y. Nakagawa, S. Tazawa, T. Wang, M. Tamura, N. Hiyoshi, K. Okumura, K. Tomishige, ACS Catal., 8, 584 (2018).
31. M. Shiramizu, F. D. Toste, Angew. Chem. Int. Ed., 51, 8082(2012).
32. M. Tamura, N. Yuasa, J. Cao, Y. Nakagawa, K. Tomishige, Angew. Chem. Int. Ed., 57, 8058 (2018).
33. J. Cao, M. Tamura, Y. Nakagawa, K. Tomishige, ACS Catal., 9, 3725 (2019).

34. R. Pinkos, R. H. Fischer, B. Breitscheidel and P. Polanek,, CA Pat, 2168458C (2004).
35. T. Wang, S. Liu, M. Tamura Y. Nakagawa, N. Hiyoshi, K. Tomishige, Green Chem., 20, 2547 (2018).
36. S. Koso, H. Watanabe, K. Okumura, Y. Nakagawa, K. Tomishige, J. Phys. Chem. C, 116, 3079 (2012).
37. T. Wang, M. Tamura, Y. Nakagawa, K. Tomishige, ChemSusChem, 12, 3615 (2019).

29th Saudi-Japanese Catalyst Symposium
November 3–4, 2019

Effects of Residual Catalyst Structure and Electronic Environment on Polyethylene Crystallization and Degradation

Muhammad Atiqullah
Senior Research Engineer, e-mail: matiq@kfupm.edu.sa
Center for Refining & Petrochemicals (CRP)
King Fahd University of Petroleum & Minerals, Dhahran 31261, Saudi Arabia

Saudi Arabia is the 3rd largest polyethylene producer in the world. Polyethylenes are one of the most important families of thermoplastics. Their innumerable applications and world-wide market make the utilization of olefins, especially ethylene, propylene, 1-butene, and 1-hexene highly profitable. Catalyst substantially contributes to making new polyethylenes (PEs) while polymer crystallization and degradation significantly affect processing and market-driven applications. Polyolefin, including PE, is a strategic research area. The virgin PE and end-products eventually and implicitly contain residual catalysts which cannot be either separated or recycled. Therefore, the present talk addresses the following challenges:

- **Q1:** Residual catalyst is an apparently **abstract** entity that exists in as-synthesized PE. Despite so, is it possible to otherwise figure out its structure and electronic environment?
- **Q2:** If answer to **Q1** is **YES**, can this help reply pending technical questions and add new insights to PE crystallization and degradation?
- **Q3:** If answer to **Q2** is **YES**, can we integrate catalyst development, PE synthesis and characterization, processing, and product development by designing an affordable and useful perspective and protocol?

With respect to what has been stated above, this talk is divided into four parts. Part I illustrates how the ***abstract*** residual catalyst structure and electronic environment can be derived using precatalyst and cocatalyst structure, catalyst surface chemistry, and polymerization and catalyst particle fracture mechanism. See Figures 1 and 2, and Table 1. A supported catalyst, unlike the unsupported analogue, fractures due to the stress (hydraulic/hydrodynamic force) exerted by the growing polymer chain. The underlying concept with applications has been patented.

Part II addresses the effects of residual catalyst structure, steric, and temperature on PE crystallinity, crystal growth and dimension, and crystallization rate. All the PEs were made under comparable polymerization conditions. Each catalyst, without a comonomer, synthesizes linear PE chain. Therefore, the afore-listed crystallization properties should be similar to comparable. But why do they differ? Why is Z–N PE is more crystalline than metallocene PE? Why does metallocene PE experience much greater crystal growth impingement than Z–N PE? Satisfactory answers are not available in the literature.

Figures 3 and 4 particularly show that dissonant residual catalyst structures and sterics variedly affect crystal nucleation and growth as well as folding of the PE chain and distribution of the residual catalyst particles in the crystalline and amorphous phases. Hence, the above phenomena occur. Model-predicted values of kinetic triplet ($n \Rightarrow$ crystal dimension (Avrami exponent), crystallization rate constant K, and $(t_{½})^{-1} \Rightarrow$ average crystallization rate) supports this argument.

Precatalysts	Ziegler-Natta	Metallocene	Post-metallocene	Late transition metal
Examples	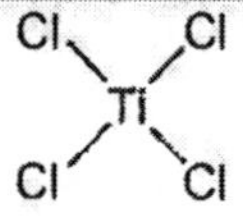	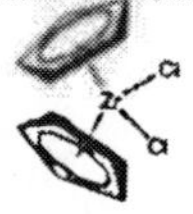	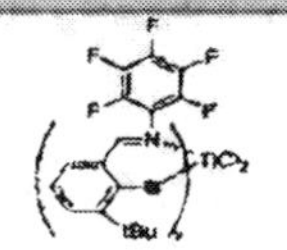	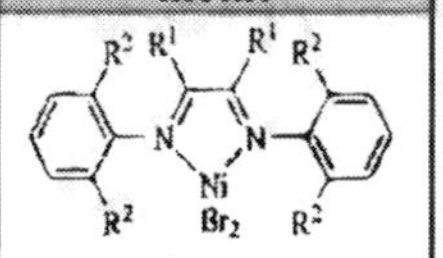

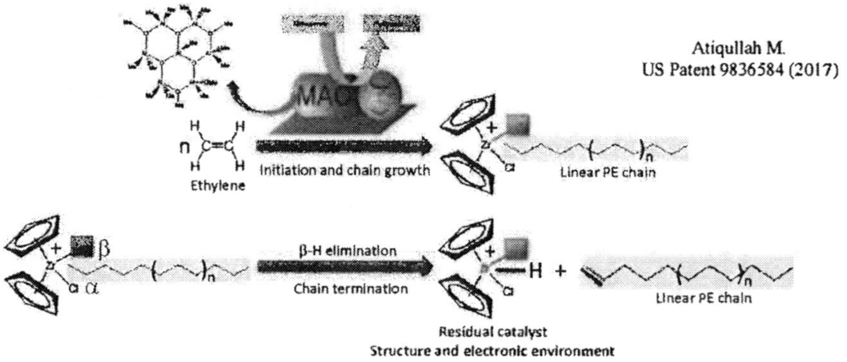

Figure 1. Derivation of residual catalyst structure and environment for unsupported catalyst.

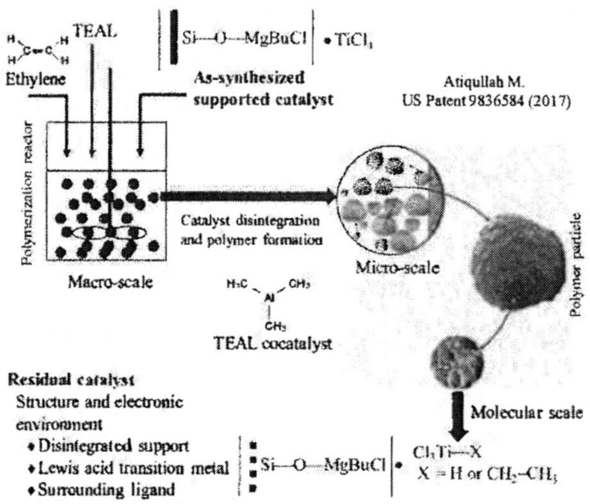

Figure 2. Derivation of residual catalyst structure and environment for supported catalyst.

Table 1. As-synthesized versus residual catalyst structure and electronic environment.

Experimental catalysts	Macro-scale as-synthesized catalysts	Molecular-scale residual catalysts
Z-N 1	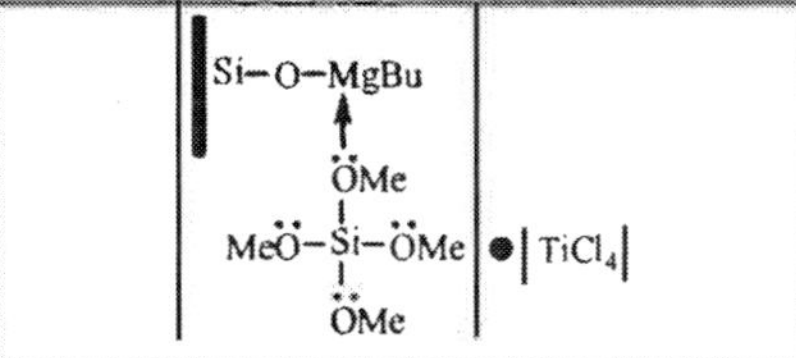	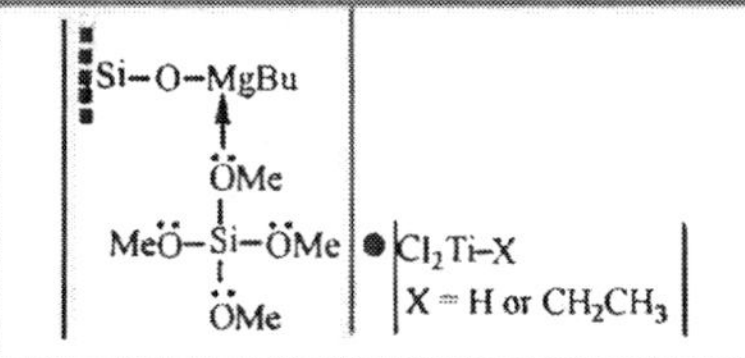
Z-N 2	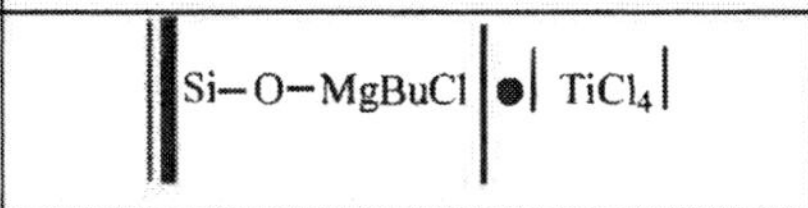	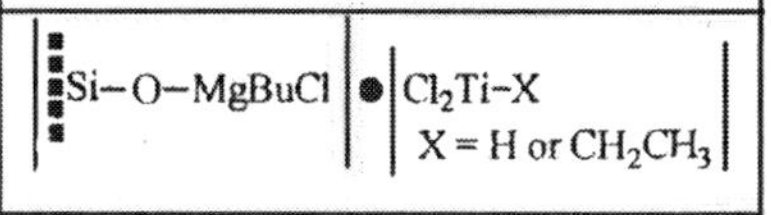
Metallocene catalyst	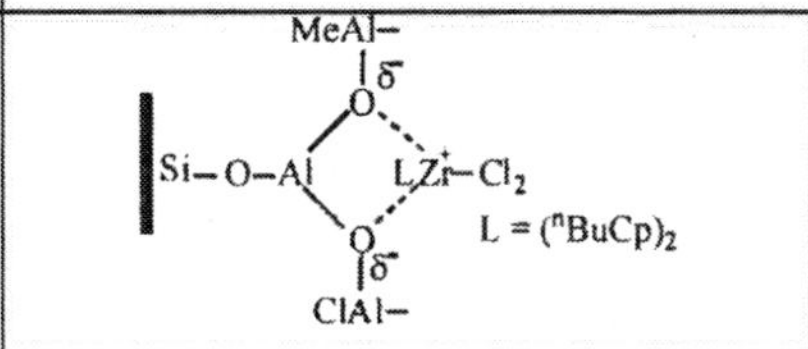	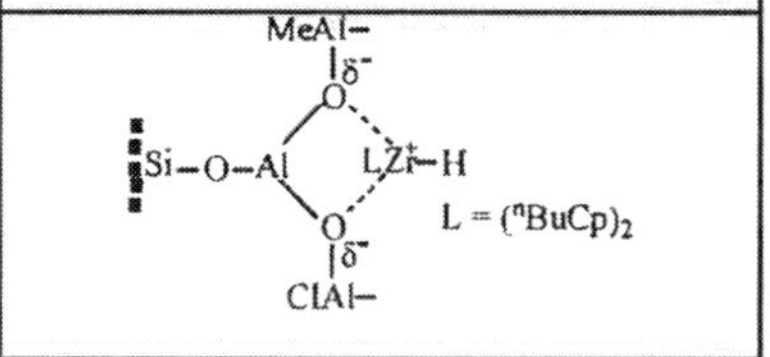

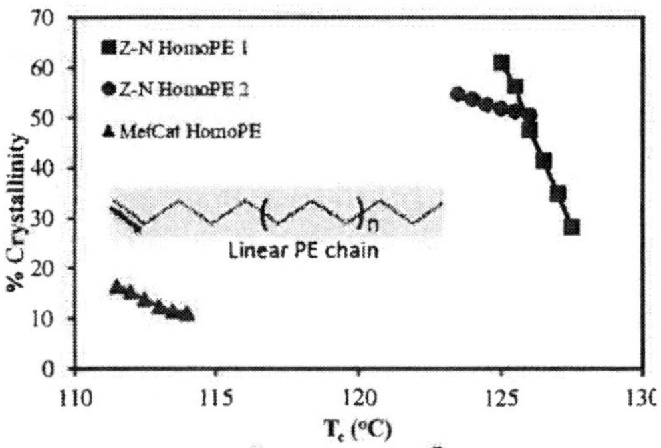

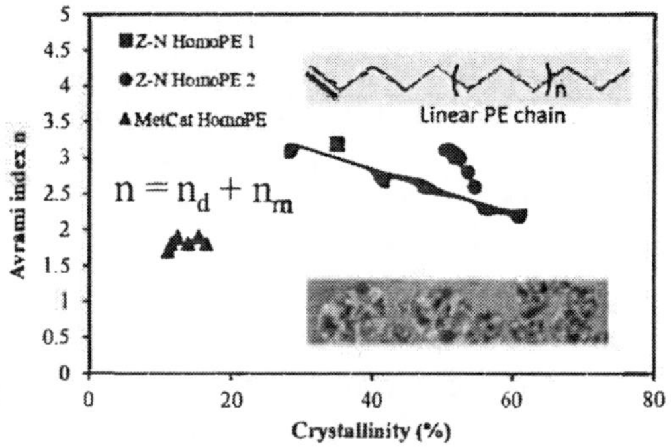

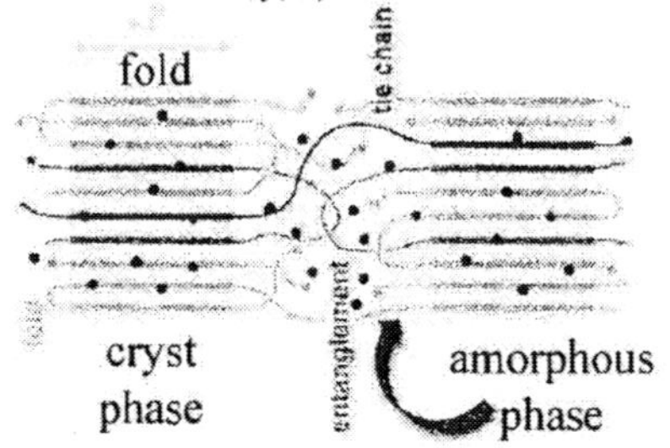

Prop	Z-N 1	Z-N 2	MetCat
M_w	402,425	180,272	115,190
T_{mp} °C	136.7	137.0	120.2
% Cryst	136.7	137.0	120.2
Tm ppm	61.0	52.4	3.5

Figure 3. Effect of residual catalyst structure and crystallization temperature on crystallinity and crystal dimension.

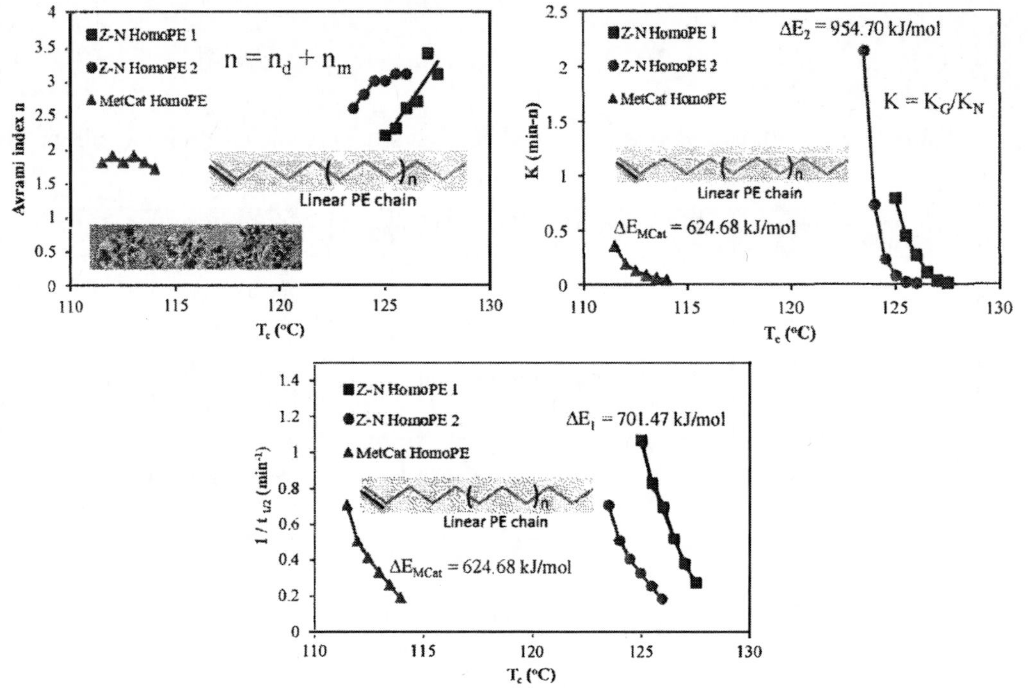

Figure 4. Effect of residual catalyst structure and crystallization temperature on crystal growth dimension and crystallization rate.

Part III underscores the effects of Z–N and metallocene residual catalysts on the UV degradation of unstabilized PE films. Figure 5 demonstrates that the metallocene film is more prone to UV degradation than the Z–N film despite the Zr level is $1/15^{th}$ of the Ti content. Why is it so? The reason is that the metallocene residual catalyst promotes the UV-mediated chain-breaking initiation reaction and decreases the hydroperoxide decomposition activation energy more than the Z–N analogues. Note that the degradation level of both Z–N films also differs. The variation in the derived residual catalyst structure, electronic environment, steric, and Lewis acidity duly explain all the aforementioned findings. The above insights, attributable to these properties and features of residual catalyst, are the first contribution to PE degradation and crystallization literature, to be credited to us.

Part IV relates to the application of residual catalyst, as defined and illustrated earlier, to develop an integrated research program that aims at making better industrial catalysts and polyolefin (PO) end-products. Figure 6 depicts the various components of this research program, and lists the challenges addressed and to be addressed.

The present residual catalyst structure and electronic environment concept equally applies to all olefin polymerization catalyst types, all kinds of PO degradation, and isothermal and nonisothermal crystallization as well as to statistical PE copolymers and polypropylenes (with stero- and rigiodefects).

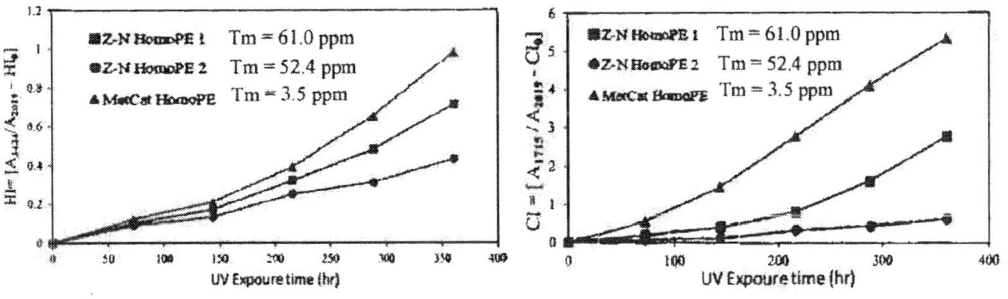

$$ROOH + ResCat^{n+} \longrightarrow ResCat^{(n+1)+} + RO\bullet + HO^-$$

$$ROOH + ResCat^{(n+1)+} \xrightarrow{UV\ h\nu} ResCat^{n+} + RO_2\bullet + HO^-$$

$$2ROOH \longrightarrow RO\bullet + RO_2\bullet + HO^-$$

Polymer Degradation and Stability, v. 95, 2289 (2010).
"The study takes this area a step further in our understanding of the role of catalyst type on oxidation rates—in this case UV, in a thorough and sound analytical approach."

Figure 5. Effect of residual catalyst structure, electronic environment, steric, and Lewis acidity on PE film UV degradation.

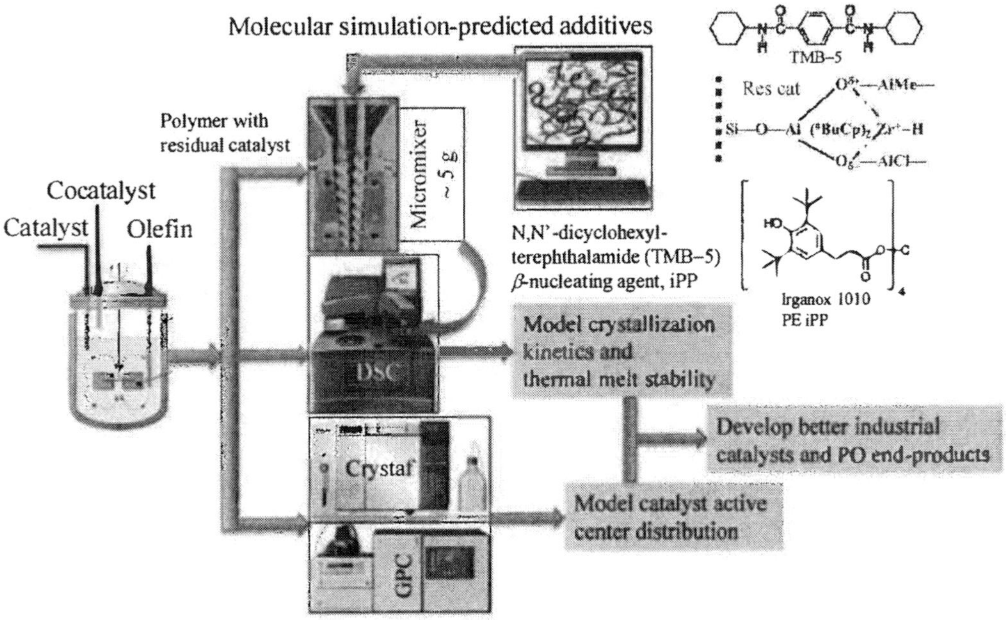

Figure 6. An integrated research program that aims at making better industrial catalysts and PO end-products: Challenges addressed and to be addressed.

Our patents and publications that prepare this contribution are listed below:

- USP 983,6584 (2017)
- USP 983,4630 (2017)
- RSC Advances, v. 7, 4249 (2017)
- Journal of the Taiwan Institute of Chemical Engineers, v. 76, 141 (2017)
- American Institute of Chemical Engineering Journal, v. 62: 1688 (2016)
- Journal of Thermal Analysis and Calorimetry, v. 119, 581 (2015)
- Polymer International, v. 63, 1824 (2014)
- Polymer International, v. 63, 955 (2014)
- Journal of the Taiwanese Institute of Chemical Engineers, v. 45, 1982 (2014)
- Journal of Polymer Research, v. 20, 56 (2013)
- American Institute of Chemical Engineering Journal, v. 59, 200 (2013)
- Industrial & Engineering Chemistry Research, v. 52, 9359 (2013)
- Current Catalysis, v. 1, 2 (2012)
- Journal of Polymer Research, v. 19, 9797 (2012)
- Polymer Degradation and Stability, v. 95, 2289 (2010)

29th Annual Saudi-Japan Symposium
November 2-3, 2019
Dhahran, Saudi Arabia

SINTERING- AND COKING- RESISTANT CORE-SHELL NANOPARTICLES IN DRY REFORMING OF METHANE

Dr. Noor Al-Mana
Research and Development Center
Saudi Aramco, Dhahran, Saudi Arabia

Abstract

Dry reforming of methane ($CH_4 + CO_2 \rightarrow 2H_2 + 2CO$) is one of the most attractive technologies to produce syngas (H_2 and CO) because it consumes greenhouse gases (CH_4 and CO_2), producing a low H_2/CO ratio that can be useful for the following, e.g., methanol synthesis or Fischer-Tropsch synthesis. The main drawback of catalysts for dry reforming is coke formation on the surfaces, which leads to reactor plugging and deactivation of the catalysts. Moreover, dry reforming of methane is endothermic reaction and thus requires high temperatures (≈900°C) to achieve high conversion, lowering the lifetime of the catalysts due to sintering of the metallic particles. The metal sintering not only causes loss of active sites, but also builds up large particles that favor formation of carbon filament. The common metals used for dry reforming are Rh, Ru, Pt, Ni and Co and they are usually supported on alumina, zirconia or silica, or others mixed oxides. Thus if these nanoparticles were trapped in a thin porous layer of oxide to form a core-shell structure, extent of sintering at high temperature may be reduced as no contact between nanoparticles would be allowed. Additionally, some oxides such as titania, zirconia, ceria or mixed oxides are known to promote carbon oxidation by activating CO_2, thus reducing coke formation during DRM reaction. Therefore, trapping metal nanoparticles in this type of oxide shell would be the way to produce catalysts that are tolerant to both sintering and coking.

To achieve the structure with protected metal nanoparticles, reverse micro-emulsion system was chosen to give metal@oxide core-shell materials. This method allows synthesizing fine and monodispersed metal nanoparticles, followed by control coating of metal nanoparticles by oxides matrix. The reverse micro-emulsion method consists of forming nano-sized droplets of water as a nanoreactor in a mixture of organic solvent (cyclohexane) and surfactant (IGEPAL CO-520). To precisely control the size and thickness of the metal@oxide nanoparticles, different parameters play a role in this synthesis, such as the surfactant and organic solvent nature, pH, water/surfactant ratio and reducing agent. Initial attempts involve synthesis of Pt@SiO_2 core-shell particles, which have been extensively studied, and then the method is extended for developing non-noble metals (Ni, Co) and other oxides (TiO_2, CeO_2, ZrO_2). The method successfully produces fine metal particles with well-structured shell using Pt and SiO_2 with controllable thickness (~15 nm). The presentation will discuss the challenges of designing metal@oxide core-shell structure with different oxides.

Oxidative dehydrogenation of propane to propylene over $VO_x/CaO\text{-}\gamma Al_2O_3$

Afees A. Ayandiran[1], Hussein K. Amusa[1], Idris A. Bakare[2], Shaikh A. Razzak[1]
Sameer Al-Ghamdi[3], Saad Al-Bogami[3], Mohammad M. Hossain[1*]

[1]Department of Chemical Engineering, [2]Center of Research Excellence in Nanotechnology, King Fahd University of Petroleum & Minerals, Dhahran, Saudi Arabia

[3]Research & Development Center, Saudi Aramco Oil Company, Dhahran, Saudi Arabia

Abstract

Novel $VO_x/CaO\text{-}\gamma Al_2O_3$ catalysts are developed for ODH of propane to propylene under gas phase oxygen free atmosphere. The catalysts are synthesized with different $CaO/\gamma Al_2O_3$ ratios keeping vanadium loading at 10 percent. The ODH of propane experiments are conducted in a fluidized CREC Riser Simulator under gas phase oxygen free conditions. The reaction temperature is varied between 550 to 650 °C at different contact time (10-30 sec). Among the evaluated catalysts, $VO_x/CaO\text{-}\gamma Al_2O_3$ (1:1) displays highest propane conversion (65 %) and propylene selectivity (85%) and the low CO_2 due to its excellent oxygen carrying capacity, balanced acidity and moderate active site-support interactions.

Keywords: Fluidized ODH, Propylene, VO_x, CaO, active site-support interaction.

*Corresponding author: Mohammad M. Hossain
Tel.: +966-13-860-1478; Fax: +966-13-860-4234.
E-mail: mhossain@kfupm.edu.sa

1. Introduction

The conventional steam cracking and catalytic cracking processes for olefin production are energy intensive and costly. Oxidative dehydrogenation (ODH) can overcome the drawbacks associated with conventional processes [Al-Ghamdi et al. 2013; Afees et al. 2016]. A metal based catalysts can play an important role achieving high propylene selectivity by minimizing the undesired CO_x formation. The present research is focused on a novel fluidized bed ODH process under gas phase oxygen free conditions (instead used lattice oxygen of the catalysts) to minimize the CO_x formation and to increase the propylene selectivity. Towards the end, $VO_x/CaO\text{-}\gamma Al_2O_3$ catalysts have been investigated as suitable catalysts for gas phase oxygen free ODH of ethane to produce ethylene. CaO was introduced to control both the acidity and oxygen carrying capacity of the catalyst.

2. Fluidized ODH of propane

The gas phase oxygen free ODH propane experiments were conducted in a fluidized CREC Riser Simulator (Fig. 1) (de Lasa 1992). Propane ODH runs were carried out at different temperature ranging from 550 °C to 640 °C while contact time was varied between 10-31 sec. The ODH of propane experiments were conducted using 0.5 g of catalyst. A preloaded syringe was used to inject 1.2 ml of feed (propane) into the reactor. At the end of the reaction period, the reactor contents were evacuated into the vacuum box. The product gas contained in the vacuum box was analyzed by gas chromatography (GC) using both a thermal conductivity detector (TCD) and a flame ionization detector (FID). Catalyst performance was studied based on propane conversion, selectivity and yield given below.

$$\text{Propane conversion, } X_{C_3H_8}(\%) = \frac{\sum_j z_j n_j}{3n_{propane} + \sum_j z_j n_j} \times 100 \quad (1)$$

$$\text{Selectivity to a product, } S_j(\%) = \frac{z_j n_j}{\sum_j z_j n_j} \times 100 \quad (2)$$

where, z_j and n_j are the number of atoms of carbon and moles of gaseous carbon containing product j, respectively. $n_{propane}$ is the mole of unconverted propane in the product stream.

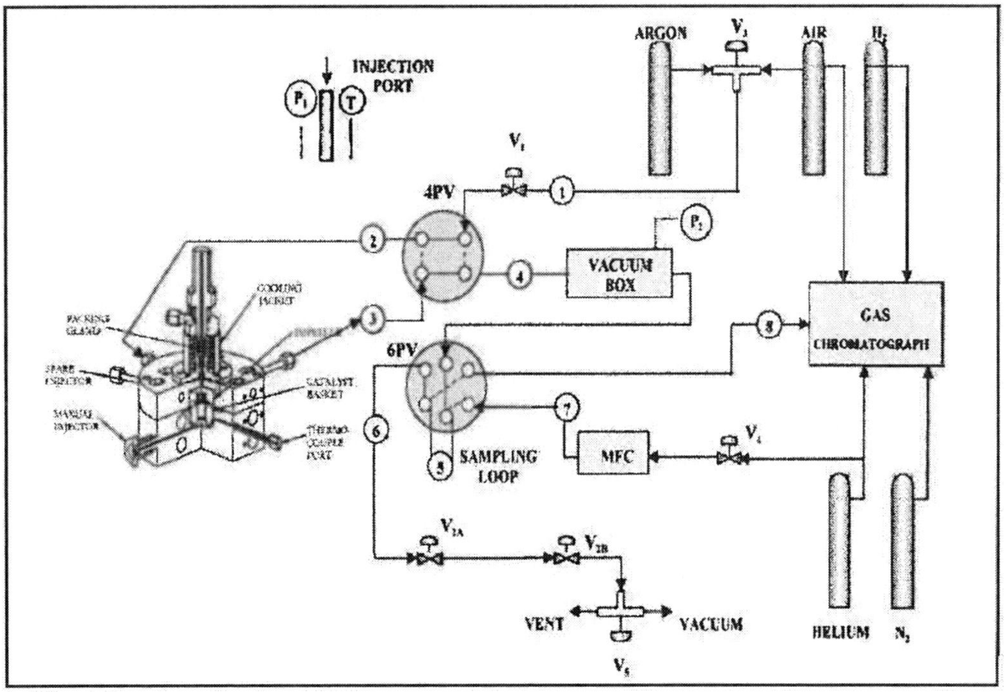

Fig. 1: Experimental set of CREC Riser Simulator (de Lasa, 1992).

3. Results and discussion

The gas phase oxygen free oxidative dehydrogenation (ODH) of propane experiments were conducted in a fluidized CREC Riser Simulator using pure propane (99.95% purity) as feed. Before performing the actual catalytic ODH runs, thermal experiments (without any catalyst) were conducted to detect any contributions from thermal conversions. The highest reaction temperature (640 °C) was selected for the thermal experiments. The GC analysis of the thermal runs products showed mainly unconverted propane and a trace amount of ethane and methane most likely due thermal cracking of propane in absence of catalyst.

In the catalytic experiments, the reaction temperature was varied between 550 and 640 °C, while the reaction was attuned from 10 to 31 sec. The product analysis of the preliminary experimental runs contain unreacted propane, propylene and carbon dioxide. Under the studied reaction conditions no hydrogen was detected, indicating the absence of cracking and/or dehydrogenation. The propane conversion and product selectivity in the experimental repeats are found to be within 2.5% error limits (Fig. 2).

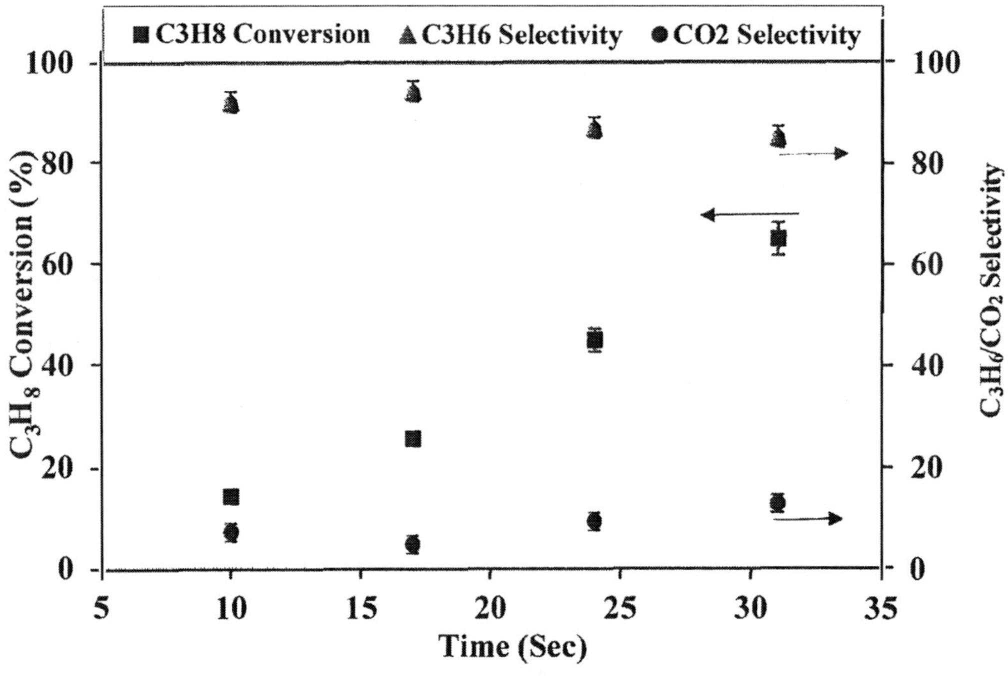

Fig. 2. Propane (C_3H_8) conversion and propylene (C_3H_6) and CO_2 selectivity and their error limits in repeated runs at T: 640 °C; Cat.: 0.5 g; Propane injected: 1.2 ml.

Mass balances were established for each of the three repeats of each individual run. Mass Balance closed consistently in excess of 95%. From the product analysis, one can consider the following possible reactions steps during the fluidized ODH of propane runs in absence of gas phase oxygen:

ODH of propane to propylene: $C_3H_8 + \frac{1}{2}V_2O_5 \longleftrightarrow C_3H_6 + H_2O + \frac{1}{2}V_2O_3$ (17)

Complete oxidation of propane: $C_3H_8 + 5V_2O_5 \longleftrightarrow 3CO_2 + 4H_2O + 5V_2O_3$ (18)

Complete oxidation of propylene: $C_3H_6 + \frac{9}{2}V_2O_5 \longleftrightarrow 3CO_2 + 3H_2O + \frac{9}{2}V_2O_3$ (19)

Therefore, it is very important to identify the best reaction conditions in order to achieve highest possible propylene yields and suppress the complete combustion reactions which produce CO_2. Keeping the above in mind the following experiments were conducted at different conditions to demonstrate the effects on the propane conversion and product selectivity of (i) the consecutive propane injection without catalyst regeneration (ii) reaction temperatures and (iii) contact times.

4. Conclusions

Following are the conclusions of the present study:

i. Gas phase oxygen free conditions favored the formation of selective product propylene and minimized the complete oxidations to CO_x. A higher degree of catalyst reductions favored the propylene selectivity.
ii. The catalyst with intermediate acidity and moderate active site-support interaction (VO_x/CaO-γAl_2O_3 (1:1)) displayed highest propylene selectivity (85 %) at higher propane conversion (65 %).

Acknowledgement

The author(s) would like to acknowledge the support provided by the Deanship of Scientific Research (DSR) at King Fahd University of Petroleum & Minerals (KFUPM) for funding this work through project No. IN161022.

References

Elbadawi, A.W.H., Khan, M.Y., Quddus, M.R., Razzak, S.A., Hossain, M.M. (2017), Kinetics of oxidative cracking of n-hexane to olefins over VOx/Ce-Al2O3 under gas phase oxygen-free environment, *AIChE J. 63, 130.*

Afees, A.A., Bakare, I.A, Binous, H., Al-Ghamdi, S., Razzak, S.A., Hossain, M.M. (2016), Oxidative dehydrogenation of propane to propylene over VOx/CaO-γAl2O3 using lattice oxygen, *Catal. Sci. Tech.*, 6, 5154.

Al-Ghamdi S., Volpe M., Hossain M.M., de Lasa I. (2013), VOx/c-Al_2O_3 Catalyst for Oxidative Dehydrogenation of Ethane to Ethylene Desorption Kinetics and Catalytic Activity, Appl. Catal., 450, 120.

de Lasa H., Riser simulator for catalytic cracking studiesm USA Patent 5,102,628 (1992)

Effect of catalyst dilution and binders on the alkylation activity of USY zeolites

Mohammed R Aldossary, Vladimir Zholobenko and Arthur Garforth

Saudi Aramco R&DC

Abstract

The growing demand for higher quality gasoline coupled with stringent environmental regulations requires the reduction in aromatics, sulfur and nitrogen compounds. The alkylation process utilizes low value olefins with isobutane producing high octane gasoline that improves the refinery's profit margin.

Current technologies utilize harmful homogenous catalysts (hydrofluoric or sulfuric acids) as catalysts for the reaction. Those acids present serious health and environmental concerns in addition with the high operational and mitigation costs. Zeolites present an ideal substitute to both acids due to the benign nature and acidic properties of zeolites. Faujasites in particular are very attractive catalysts for the alkylation reaction due to their large-pores and high acidity which can be tailored towards alkylation. Deactivation remains a major challenge for a zeolite-based alkylation process.

The effect catalyst dilution by inert binders, namely: alumina, silica, MCM-41 and SBA-15, on the activity, stability and selectivity of the faujasite USY was studied using a well-mixed, continuous stationary Berty-type reactor at 75 °C, 17 bar, an olefin space velocity of 0.30 h^{-1} and an isobutane/2-butene ratio of 15 (mol/mol). The dilution significantly improved the catalytic stability with the diluted zeolites reaching higher butene conversions by 35 % after 10 hours on-stream mainly due to the availability of large-mesoporous inert channels which facilitated the removal of the products and prevented further hydride transfer with alkylate products. The optimum diluent was MCM-41 due to its high surface area. Varying the inert material had minor improvements after the initial addition.3

Heavy naphtha reforming of C12 over variable-morphological MFI zeolites

Mohammed A. Sanhoob[1,2], , Emad N. Shafei[3], Mohammad Z. Al-Bahhar[3], Ali N, Al-Jishi[3],

Ki-Hyouk Choi[3], Abuzar Khan[1], Oki Muraza*[1,2]

[1]Center of Excellence in Nanotechnology, King Fahd University of Petroleum & Minerals (KFUPM), Dhahran 31261, Saudi Arabia.
[2]Chemical Engineering, King Fahd University of Petroleum & Minerals (KFUPM), Dhahran 31261, Saudi Arabia.
[3]Research and Development Center, Saudi Aramco, Dhahran 31311, Saudi Arabia.

Abstract

MFI zeolite with different silicon sources was synthesized hydrothermally. The study investigated the effects of silicon source on particle size and zeolitic morphology in order to determine the reaction stability for naphtha reforming. In the study, adjusting zeolitic particles sizes, adaptation of synthesis conditions such as the synthesis time and hydrothermal temperature were applied. With an increase in the synthesis temperature, particle morphology thickness became thinner and smaller with various isotherm types I and IV as a dominant textural property. Due to the morphological and particle sizes variations, the concentration of both Lewis and Brønsted acids sites were also varied. The Brønsted acid site strength was affected by the types of silica source. The study emphasized the importance of the catalytic reactivity, stability, and selectivity to be correlated with zeolitic properties as a toolbox to improve the zeolitic performance for heavy naphtha reforming for more cyclization and transalkylation reactions.

Keywords

Variable-size MFI; silica source; high conversion; naphtha reforming.

*Corresponding Author (OM)
E-mail: omuraza@kfupm.edu.sa, Phone: +966 13 860 7612

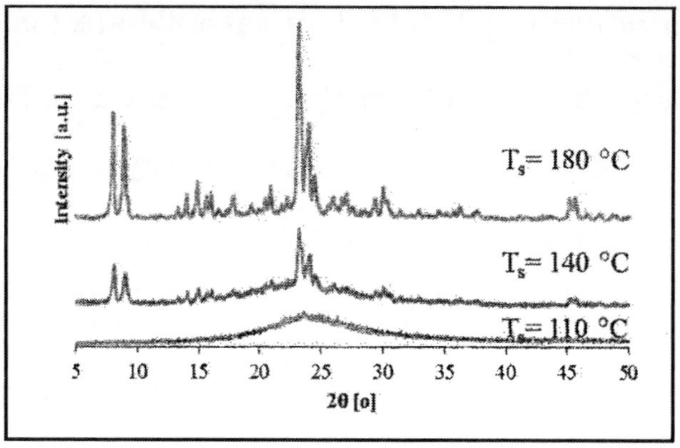

Figure 1. XRD patterns of modified MFI zeolite synthesized under hydrothermal conditions of tetraethyl orthosilicate (different synthesis temperatures, Ts).

New Generation Family of Higher Performance Reforming Catalysts

Ali Jahel

Axens

Abstract

Symphony® is a harmonious orchestration of 100 years of cumulative reforming catalyst expertise scored to yield superior performance and client satisfaction. Axens' acquisition of Criterion's reforming business has allowed for unprecedented synergies between superior catalyst support technologies and superior multi-metal formulation technologies. The results of these synergies are reflected in compelling improvement in yields and hydrothermal stability for virtually all reforming services from:

• Fixed bed to CCR
• Low to high density loading
• Lean to rich naphtha feed
• High to low octane severity

Symphony® should be the catalyst of choice with replacement profitability so conclusive in cases that even a change-out before end-of-life assures payback in months rather than years.

The benefits offered by Symphony® catalysts for CCR reforming applications can be visualized for example when new PS 100 and PS 110 references are benchmarked to the well-known, previous best in class generation of CCR reforming catalysts, PS 40. These benefits are achieved without sacrificing catalyst activity. Benefits include improved hydrothermal stability as well as Improved C5+, aromatics and H_2 yields across the full spectrum of feed quality and product octane.

For example, C5+ yield is up to 1.5 wt% higher with Symphony® PS 100 versus PS 40 when Arab Light naphtha is reformed to 100 RON. Consequently, aromatics and H_2 yields improve respectively, for the same feed and octane severity. C5+ yield improvement also remains constant over a wide range of octane severity (94 101 RON) with Symphony® PS 100 versus PS 40 when typical European naphtha (100°C-160°C) is reformed. Aromatics and H_2 yield improvements remain constant for a wide range of octane. All of the above-mentioned benefits are achieved without sacrificing catalyst activity.

Furthermore, throughout their lifetime, new symphony references offer higher surface area than the previous generations as in the case of PS 100 versus PS 40. As a result, new Symphony® catalysts take over twice as long to attain the same surface area.

This affords superior retention of requisite acid and metal sites for optimized catalyst performance. On the other hand, new Symphony® catalysts offer superior reliability through less organic chloride addition and consequent lower downstream equipment corrosion & fouling.

This presentation will go through the new generations of high and low density CCR catalysts from the Symphony® series, demonstrating the long innovation journey of Axens in the field of naphtha reforming catalyst development and rapid introduction to the market place.

The Effect of Zeolite Crystal Structure on Hydrocarbon Conversion

Mohammad Al-Bahhar[1], Emad N. Al-Shafei[1], Ali N. Al-Jishi[1], Oki Muraza[2], Mohammed A. Sanhoob[2], Anas K. Jamil[2], Ki-Hyouk Choi[1], Ali S. Alnaser[1] and Hameed H. Al-Badairy[1]

[1] Research and Development Center, Saudi Aramco, Dhahran 31311, Saudi Arabia

[2] Center of Excellence in Nanotechnology and Chemical Engineering, King Fahd University of Petroleum & Minerals (KFUPM), Dhahran 31261, Saudi Arabia.

Abstract:

Zeolites are one of the major components of catalysts in refinery and petrochemical industry. In spite of well-established protocols for using zeolite in commercial application, there are still several challenges that limit its performance with regards to selectivity and catalytic reactivity. Activity of a zeolite is generally deteriorated by coking and alumina framework oxidation state changed during reaction. This study aimed to investigate different crystal zeolite structure with better performance for hydrocarbon conversion reaction [1-2]. In this study, zeolite crystal size was controlled to reach nanoscale crystals to reduce the diffusion path length for improving and accelerate the reaction rate of primary and secondary reactions of hydrocarbon. The effects of synthesis parameters on the catalyst reactivity and hydrocarbon conversion stability of nanocrystal MFI and TON zeolites was investigated. The primary and secondary cracking of hydrocarbon product yield was found to strongly depend on the stability of the framework and zeolite dimensional intracrystalline pores structure. Nano-structured MFI zeolite showed much higher and more stable conversion in hydrocarbon cracking over time than that of conventional micro MFI. This study also investigated the reaction pathway of hydrocarbon cracking, cyclisation and isomerization to optimize the yield of reformate naphtha product. It was concluded that short diffusion path and 3D structural pores of nano zeolites were significant factors in naphtha reforming to gasoline.

Reference:

1. Anas K. Jamil, Oki Muraza, Ryota Osuga, Emad N. Shafei, Ki-Hyouk Choi, Zain H. Yamani, Ali Somali, and Toshiyuki Yokoi (2016), Hydrothermal stability of one-dimensional pore ZSM-22 zeolite in hot water, *J. Phys. Chem.* 120, 40
2. Mohammed A. Sanhoob, Oki Muraza, Emad N. Shafei, Toshiyuki Yokoi, Ki-Hyouk Choi, Steam catalytic cracking of heavy naphtha (C12) to high octane naphtha over B-MFI zeolite, Applied Catalysis B: Environmental, Volume 210 (2017) 432-443.

Machine Learning Approach for Prediction of Reaction Yield **with Simulated Catalyst Parameters**

Akira Yada
Interdisciplinary Research Center for Catalytic Chemistry
National Institute of Advanced Industrial Science and Technology (AIST)
1-1-1 Higashi, Tsukuba, Ibaraki 305-8565 Japan
E-mail: a-yada@aist.go.jp

ABSTRACT

Rational design and optimization for the development of a highly reactive catalyst requires the synthesis of various catalysts and the examination of several reaction conditions, however this traditional "trial-and-error" approach generally costs time and labor. The construction and application of a predictive model for the catalyst performance with informatic techniques such as the machine learning would facilitate the process of catalyst development, allowing more rapidly to approach the highly reactive catalyst. This "catalyst informatics" approach has attracted much attention in various chemical industries such as functional material, pharmaceutical, and petrochemical industries, being expected to accelerate the development of more efficient catalysts and industrial processes. In this presentation, the recent progress in this area such as the prediction of the catalytic reactivity and the optimization of the reaction conditions with artificial intelligence (AI) will be introduced. In addition, our recent result in the machine learning approach for the prediction of the catalytic reaction yield, which is aiming at the discovery of innovative catalyst by AI, will be presented.

1. Introduction

Discovery and development of new catalyst is of great importance, because, whenever an innovative catalyst is discovered, scientific innovation take place and our social life is enriched. At present, many chemical transformations are relying on the power of catalysis. In fact, various catalysts are widely utilized in industrial processes such as petroleum refining and petrochemistry, manufacturing for pharmaceuticals, agricultural chemicals and functional materials, environmental cleanup, and resource and energy saving processes. The development of practical catalysts applied to the above-mentioned processes generally requires a huge number of trial-and-error essays. For example, in the development of the catalyst for the Haber-Bosch process, performed more than 100 years ago, more than 20,000 experiments were carried out for about 2,500 metal catalysts, and it took more than 10 years to complete the development of the practical catalyst [1]. Although this is an exaggerated example, even today the catalyst development requires such trial-and-error research processes. In addition, novel innovative ideas and intuition from the chemist's knowledge and experience are also involved. Furthermore, observation and analysis skills that do not overlook small discoveries, as well as interpretation of the results to imagine and design the new catalysts are also necessary. Serendipity and even mistake sometimes bring successful results. In any case, all innovative catalysts developed so far are the results of the constant efforts by many researchers.

Despite the availability of various practical catalysts, the importance of new catalyst development is increasing more and more. In fact, various environmental issues such as global warming and material recycling are expected to be solved with innovative catalysts and molecular transformations. The development of such innovative catalysts, however, must be necessary to spend time, and to cost money and labor, despite significant progress in experimental techniques, analytical instrument, and so on compared with those in 100 years ago. Furthermore, it is not difficult to imagine that novel ideas for the design of innovative catalyst are decreasing with the time. For these reasons, the construction of new methodologies that accelerate the discovery and development of innovative catalysts and thus make breakthroughs in catalytic research is highly desired.

2. Catalyst informatics

Recently, the fourth science, data science, has attracted much attention in fundamental research. In particular, artificial intelligence (AI) is promising and has been applied and implemented in various area such as biology (bioinformatics and cheminformatics), pharmaceuticals (drug discovery informatics), material science (materials informatics). We have also paid attention to the application of AI technique such as machine learning in catalyst research, and launched "catalyst informatics" since 2014. Catalyst informatics is the concept that integrates the catalyst research and data science (informatics) [2,3]. The research is directed toward the discovery of new knowledge in catalysis and the creation of completely new highly active catalysts. It is difficult for human to deal with a huge amount of data that has been published in patents and papers, or unpublished and stored in the laboratory in the universities and company institutes. Using data science methods, it is expected to gain the new knowledge of catalysts from the big data. This will not only accelerate the research and development, but also discover the new theories and principles.

Catalyst development with the machine learning approach is performed following general scheme (Figure 1). First, a prediction model is constructed to fit the reaction results such as the yield and selectivity with catalyst used in the reaction. For this purpose, catalytic reaction data is collected from the literatures and patents, or by carrying out catalytic experiments. Next, for the construction of the prediction model, parameters describing the nature of catalyst are necessary as input for machine learning. Physical parameters (e.g. molecular weight, boiling point, heat capacity, etc.) of the catalyst and calculated parameters by computational chemistry that quantifies the structure and other characteristics are collected. Then, machine learning is performed in order to optimize a prediction model that fits the reaction result (output) with the catalyst parameters (input). With this trained model, the reaction result of an unverified catalyst can be predicted by inputting the catalyst parameters into the model. In the actual catalyst development, the parameters of promising catalysts that are designed by chemist are collected, and the catalyst candidates are narrowed down with the prediction model. The selected catalysts are synthesized and their activity is evaluated. If the catalyst activity is high enough, the development will be finished. If not, however, the evaluated catalyst data is added and the prediction model is reconstructed to improve the predictability. It is expected that by repeating this process highly active catalyst might be discovered very quickly.

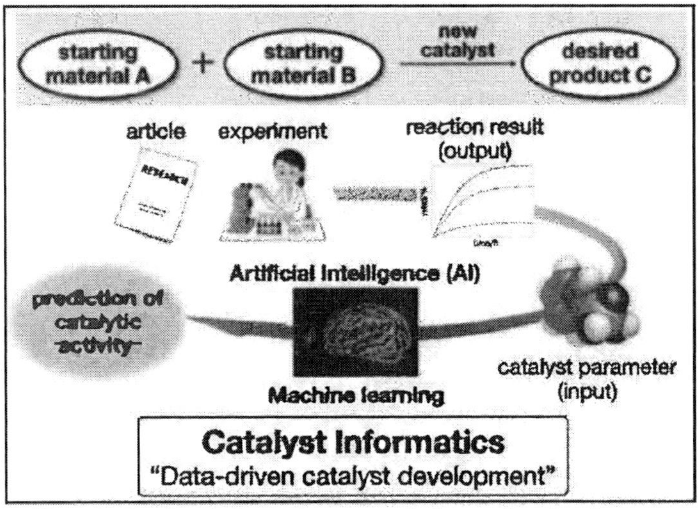

Figure1. Catalyst informatics approach for catalyst development

3. Machine learning approach for the prediction of the catalytic reaction yields [4]

In 2018, we succeeded in developing a method to construct the prediction model for catalytic reaction yields by machine learning. With this trained model, the yields of unverified catalysts were predicted. Details are described below.

In this study, the tungsten-catalyzed epoxidation reaction of olefin with hydrogen peroxide [5], which was originally developed by Sato and Noyori and has been applied to material syntheses at AIST, was selected as a model reaction. This reaction provides epoxides without contamination of any halogen atom, which are utilized in various functional organic materials such as electric materials, resin, pharmaceuticals, agrochemicals and so on. Because of the formation of H_2O as the sole byproduct, this reaction is a clean oxidation technique. The reaction requires three catalytic components, sodium tungstate dihydrate, methyltri-*n*-octylammonium hydrogen sulfate as a phase transfer catalyst, and phosphonic acid as an additive. Among them, phosphonic acid is crucial for the high reactivity. To construct a regression model that predicts their catalytic performance, experimental results (catalytic reaction yields) examining various phosphonic acids from published or unpublished data were collected as the output for machine learning. As depicted in Figure 2, the epoxide yields depend on their molecular structures, which suggests that molecular parameters describing the nature of the phosphonic acid would correlate with the reaction yield. Molecular parameters of the phosphonic acids were then collected by density functional theory (DFT) calculation and used as the input for machine learning. The calculated molecular parameters are listed in Figure 3. For example, HOMO-LUMO gap, natural bond orbital (NBO) charge, as well as frequencies, intensities, and fluctuation of the IR modes including phosphonyl group [-P(=O)(OH)$_2$] vibrations were collected (30 parameters in total).

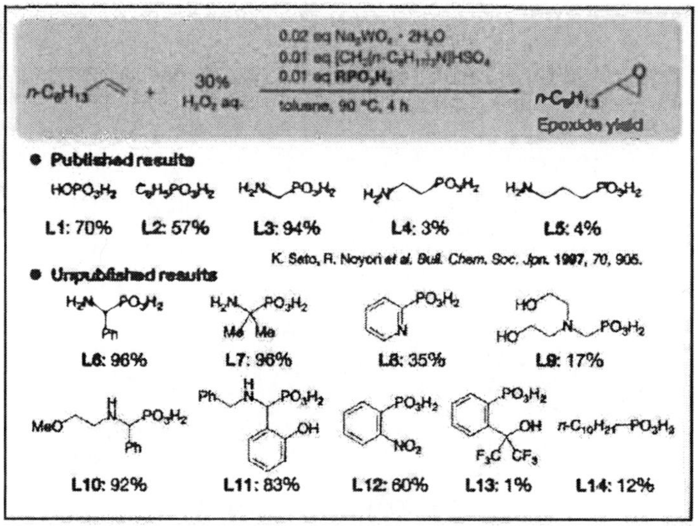

Figure 2. Scope of phosphonic acids in the tungsten-catalyzed epoxidation with hydrogen peroxide.

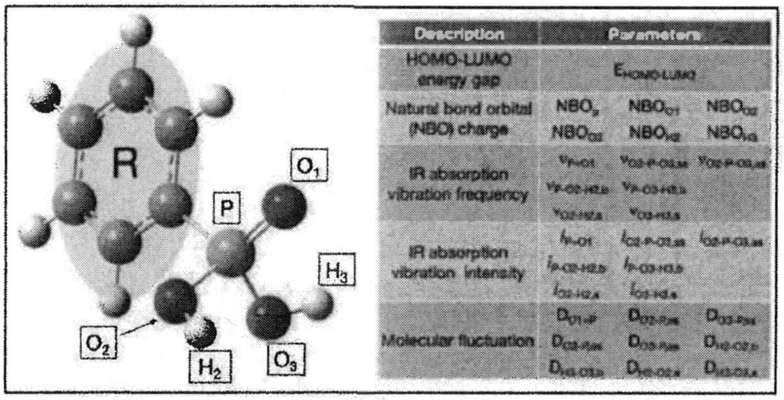

Figure 3. Summary of collected parameters for phosphonic acids.

With the catalyst parameters in hand, we next examined the regression analysis to construct the prediction model. For this regression analysis, a logistic regression model was used as shown in equation 1:

$$\text{yield}(\%) = \frac{1}{1+\exp(-t)} \times 100 \qquad (1)$$

$$t = a_0 + a_1x_1 + \cdots + a_ix_i + \cdots + a_{30}x_{30}$$

where, x_i, a_0, and a_i represent parameters of phosphonic acid molecules, the intercept, and their coefficient, respectively. For the machine learning method, LASSO (Least Absolute Shrinkage and Selection Operator) [6] was adapted. With LASSO, even when the number of parameters is larger than that of experimental data, valuable parameters for predicting reaction yields are automatically selected and a regression model is optimized, avoiding overfitting. Therefore, LASSO was performed, and the formulated equation is shown in Figure 4(a). For the present epoxidation reaction, 11 parameters were required to optimize the prediction model. The calculated yields using this predictive model and the experimental yields are plotted in Figure 4(b). The root mean square error (RMSE) value, which is an index of prediction accuracy for the data, was 5%.

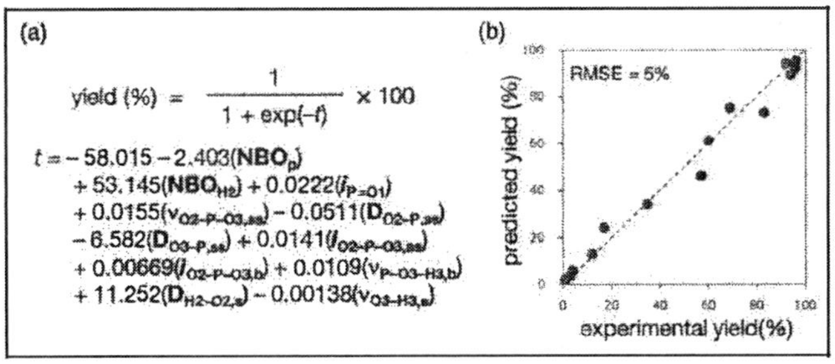

Figure 4. (a) Formulated prediction model by LASSO. (b) Experimental versus predicted yields plot for training data set.

The developed model can also be applied to predict the reaction yields using unverified catalysts. The predicted yields were calculated by entering the calculated parameters of each unverified catalyst into the predictive model, while the catalysts were also subjected to the epoxidation reaction to evaluate the prediction performance (Figure 5(a)). As shown in the experimental versus predicted yields plot (Figure 5(b)), a good agreement of the experimental yields with predicted yield was observed specially for the catalysts giving high and low yields, whereas the predicted yields of the catalysts giving moderate experimental yields were found to be outliers. Although the RMSE value for the test data was 26%, this is the first example where reaction yields could be predicted with a regression model that is developed through the automatic selection of the valuable parameters by machine learning (LASSO). In addition, this prediction model will provide us valuable information about the reactivity trend of unverified catalysts, which would be helpful for the selection of the priority order for the catalyst screening experiment.

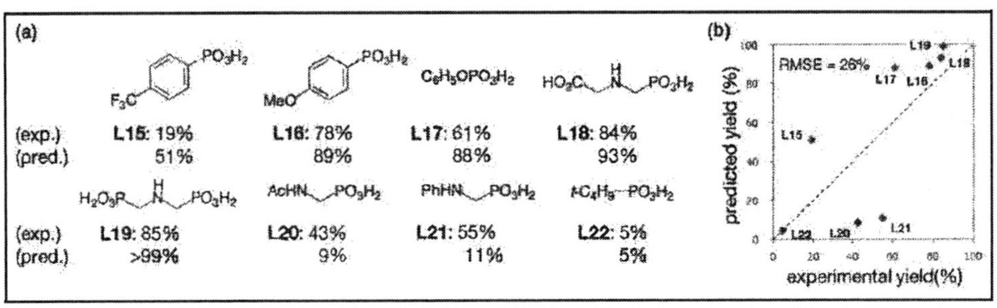

Figure 5. (a) Experimental and predicted yields with unverified phosphonic acids. (b) Experimental versus predicted yields plot for test data set.

By evaluating the standard partial regression coefficients β for the selected parameters, analysis of the contribution of each parameter is possible (Figure 6). Thus, the predictive model has a higher degree of dependence on a specific parameter having a larger β value. For instance, the infrared (IR) absorption intensity of the P=O1 stretching vibration ($i_{P\text{-}O1}$) parameter was identified as the most important one, leading the increase of the reaction yields. In contrast, the parameters having larger negative β value such as $D_{O3\text{-}P,ss}$ parameter decrease the reaction yields. In such a manner, interpretation of the β value might give valuable information for the future design of new catalysts.

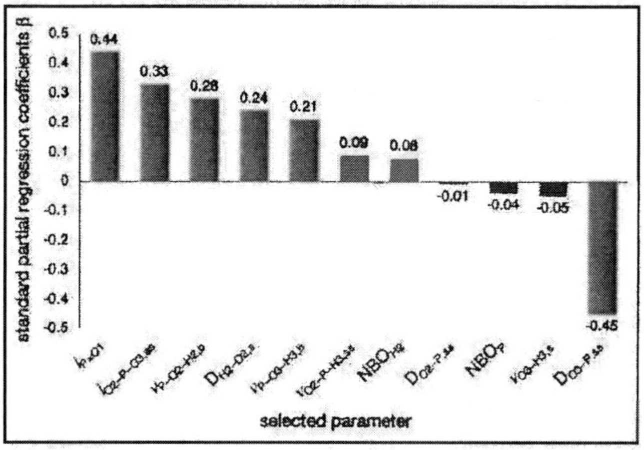

Figure 6. Values of standardized partial regression coefficients β.

4. Conclusion

In summary, the machine learning approach for the prediction of the catalytic reaction yields was demonstrated. Although the machine learning approach described here is an example of a homogeneous catalyst system, the same method can also be applied to heterogeneous catalyst systems [7,8]. Catalyst informatics is a promising tool for the future development of innovative catalysts because the method provides valuable information for the optimization of more reactive catalyst. Currently, catalyst informatics research is being investigated by many researchers. Successful examples of the utilization of machine learning for the analysis of large amounts of experimental data obtained by high-throughput experiments have been demonstrated [9,10]. However, it is still in the initial stage for the development of predictive models for the small amounts of data. The improvement of the method for developing a robust model for such data as well as the application to various catalytic reactions is ongoing in our laboratory.

Acknowledgements

This work is partly based on results obtained from a project (P16010) commissioned by the New Energy and Industrial Technology Development Organization (NEDO).

5. References

[1] M. Appl, *Ullmanns Encyclopedia of Industrial Chemistry*, **2011**, DOI:10.1002/14356007.o02_o11.

[2] K. Takahashi, L. Takahashi, I. Miyazato, J. Fujima, Y. Tanaka, T. Uno, H. Satoh, K. Ohno, M. Nishida, K. Hirai, J. Ohyama, T. N. Nguyen, S. Nishimura, T. Taniike, *ChemCatChem*, **2019**, *11*, 1146.

[3] A. J. Medford, M. R. Kunz, S. M. Ewing, T. Borders, R. Fushimi, *ACS Catal.* **2018**, *8*, 7403.

[4] A. Yada, K. Nagata, Y. Ando, T. Matsumura, S. Ichinoseki, K. Sato, *Chem. Lett.* **2018**, *47*, 284.

[5] K. Sato, M. Aoki, M. Ogawa, T. Hashimoto, R. Noyori, *J. Org. Chem.* **1996**, *61*, 8310.

[6] R. Tibshirani, *J. R. Stat. Soc. B* **1996**, *58*, 267.

[7] B. R. Goldsmith, J. Esterhuizen, J.-X. Liu, C. J. Bartel, C. Sutton, *AIChE J.* **2018**, *64*, 2311

[8] P. S. Lamoureux, K. T. Winther, J. A. G. Torres, V. Streibel, M. Zhao, M. Bajdich, F. Abild-Pedersen, T. Bligaard, *ChemCatChem*, **2019**, *11*, 3581.

[9] D. T. Ahneman, J. G. Estrada, S. Lin, S. D. Dreher, A. G. Doyle, *Science*, **2018**, *360*, 186.

[10] J. M. Granda, L. Donina, V. Dragone, D.-L. Long, L. Cronin, *Nature*, **2018**, *559*, 377.

Advancing the Frontier in Shape Selective Catalysis to Enable the Production of Petrochemical Feedstocks in the Modern Refinery

E. Vogt, G. Yaluris, C. Pouwels

Catalyst manufacturers to enable the production of increased amounts of propylene, butylene, and possibly in the future aromatics from the FCC unit, are utilizing shape selective zeolite technologies. Among these zeolites, ZSM-5 and the stabilization needed to survive in the demanding environment of the FCC unit, have emerged as key technologies. It is therefore critical to understand how the well-known ZSM-5 structure is modified by the phosphorous used in the stabilization process. We have investigated the average and local structure of phosphorus-treated HZSM-5 zeolite by a number of different techniques such as atom probe tomography, X-ray diffraction (at ambient and cryogenic temperatures) and 1H, 29Si, 27Al, 31P magic angle spinning (MAS) solid state nuclear magnetic resonance (NMR) spectroscopy. Phosphation leads to breaking of the Si-OH-Al bridging groups, and consequent partial dealumination of the framework. Most of the framework Al is present in an electronic environment distorted by the presence of phosphorus and appears not to be involved in classic Si-OH-Al Brønsted acid sites. The 31P MAS NMR indicates that phosphorus interacts with the zeolitic framework to locally form silico-aluminophosphate (SAPO) domains and the presence of a new kind of acidic site is confirmed. Increasing the phosphorus loading promotes further dealumination of the framework and cross-dehydroxylation between P-OH and Si-OH species, leading to the formation of a crystalline silicon orthophosphate phase, deactivating the zeolite. This new understanding on how phosphorus interacts with ZSM-5 and of the optimum where maximum stabilization occurs has enabled Albemarle to develop pioneering shape selective catalytic technologies. AFX™ has the selectivity and stability to be successful in the most demanding applications, such as the recent very successful new unit start up in NSRP. They also underlie the development and successful demonstration in FCC units of our new DURAZOOM™ HA olefins additive. In these new technologies, we have been able to maintain the efficiency and stability of the ZSM-5 zeolite, packing increasing performance in the same particle, and allowing refiners to achieve their increasingly challenging yield objectives.

Upgrading oil using a non-conventional charged droplets approach

Dr. Chanbasha Basheer,

Department of Chemistry, King Fahd University of Petroleum and Minerals, Dhahran, Saudi Arabia

The conversion of heavy oil fractions into lighter oil fractions is highly essential for meeting the growing demand for fuels in continuous depletion of oil reserves and for environmental remediation. we are proposing a charged-droplet assisted flow reactor for the catalytic breakdown (redox reactions) for selected petroleum products which includes asphaltene and heavy crude oil fractions based on the preliminary study, we have demonstrated for the nanoparticle-assisted degradation of asphaltene in charged microdroplets. A charged-droplet assisted flow reactor will create more opportunity to change the way petroleum chemistry performed both at a research and industrial. We selected two model polycyclic aromatic hydrocarbons as well as a crude oil sample show degradation on the microsecond time scale. Heterogeneous catalysis of a biphasic (organic/aqueous) reaction occurs at the interface of microdroplets and the charged TiO_2 surface. The fragmentation yield is increased by cycling the process. The relative fragmentation yield of a model compound (rubrene) increased from 16.9% in the first cycle to 32.6% in the third cycle. This convenient and efficient method suggests possible future industrial applications.

Catalytic Cracking of Pentene to Propene and Ethene over ZSM-5 Zeolite

Abeer Arjah [1]

[1] Research and Development Center, Saudi Aramco, Dhahran, Saudi Arabia

Abstract

Ethene and propene are very important raw materials that are used as building blocks for manufacturing solvents, fibers and plastics [1,2]. Due to the high demand of these important raw materials and the shortage in their supply it is imperative to find more ways of producing them in order to close the gap between the supply and demand [3].

Ethene and propene are most commonly produced from steam cracking and FCC consecutively. They can also be formed by converting more abundant less valuable materials in the FCC process, such as pentene, to more valuable ethene and propene via cracking [4]. Research has shown that catalytic cracking using an acidic mesoporous ZSM-5 catalyst was among the best ethene and propene production processes.

This project investigated the cracking of pentene over an acidic ZSM-5 catalyst in a fixed bed micro reactor. In order to maximize selectivity towards ethene and propene a variety of operating conditions were tested which included different reaction temperatures (450, 500 and 550°C). It was found that higher reaction temperatures were more beneficial and the best reaction temperature was 550°C. The best operating conditions were feed concentration of 2.8 mol% and WHSV =1.9 h^{-1}.

The catalyst shape selectivity was also investigated by testing different sized ZSM-5 crystals that were manufactured in-house. A range pf crystal sizes between 0.2 and 100 μm were tested. The highest conversion of 99 mol%, the highest C_2= selectivity of 46 mol% and the highest C_3= selectivity of 43 mol% were achieved at a crystal size of 65 μm.

Biography:

[1] Song, C. and Z. Yan, Characterization and Cracking Performance of the Modified USY Zeolites with Mixed Organic Acids. Fuel Chem., 2004. 49(1): p.26-27.

[2] Speight, J.G., The Chemistry and Technology of Petroleum. 4th. 2006, Boca Raton, Florida: CRC Press.

[3] Vogt, E.T.C. and B.M. Weckhuysen, Fluid Catalytic Cracking: Recent Developments on the grand old lady of zeolite catalysis. Chemical Society review, 2015. 44 (20): p. 7342-7370.

[4] Sadeghbeigi, R., Fluid Catalytic Cracking Handbook. 1995: Elsevier.

Application of hierarchical zeolites in FCC catalyst

Shakeel Ahmed[1], Adeel Ahmad[2] and Abdallah Al-Shammari[2]

[1]Center for Refining & Petrochemicals, Research Institute, King Fahd University of Petroleum & Minerals, Dhahran-31261, Saudi Arabia.
[2]Chemical Engineering Department

Abstract

With the depletion of crude oil reserves, more focus have been shifted towards the catalytic cracking of heavier hydrocarbons for the production of transportation fuels. Fluidized catalytic cracking (FCC) of vacuum gas oil (VGO) is considered as a promising process for enhancing the gasoline yield to fulfill the global energy demand. Recent literature on FCC catalyst indicates the room for increasing the efficiency of regular FCC zeolites based catalysts. This is due to the complex feed compositions, diffusion limitations, less yield of desired hydrocarbons and instability of the catalyst due to the coking and poisoning by the sulphur, nitrogen and oxygen bearing species in the feedstock. The inclusion of hierarchical zeolite catalyst systems have been considered as most suitable option. Although, it has been proved that FCC catalyst mesoporosity enhances the catalyst performance but it does not mean that ultra large pore size of zeolite are necessarily suitable active component of FCC catalysts. Both MCM-41 and USY zeolites have been tested for the large molecules cracking but MCM-41 performs poorer due to its low thermal stability at the high temperature which prevents its use in severe FCC cracking process conditions. Apart from ZSM-5 and Y zeolite, Beta zeolite has also been tested for the FCC cracking but thermal stability and economics impedes its application to use it in large scale. A catalyst mixture of ZSM-5 and ITQ-33 had more yield of middle distillate and propylene as compared to US-Y Zeolite but stability and economics limits its widespread applications. These new trends of catalyst made possible to process heavier crude oil fractions to increasing amount of propylene and gasoline range fuels.

1 Introduction

Fluid catalytic cracking (FCC) is a process which is used in oil and gas industries as a conversion technology to produce major part of the gasoline. One important thing is to note that FCC is not only the single process used in refineries for the conversion but hydrocracking is also used for the same purposes. Even though, Fluid catalytic field has been studied already for 75 years but still this area is central as a research point of view for both the industry and academia[1].

In FCC, the choice of a most suitable catalyst is very crucial for the success of a process as a result of high gasoline yield. Although, there are different zeolite based catalysts have been evaluated but less yield of the desired products, huge energy requirements for the process and poor stability of the catalyst at high temperature leads to the search of the better suitable catalyst design options [2-4]. Y-zeolite in many improved forms is using as the main cracking component of the FCC catalyst since 1964[5]. Zeolite crystals with some unique characteristics such as shape selectivity, high surface area and acidity had wide applications in industry as a heterogeneous catalyst. Due to their small pore size <1.5nm, zeolite crystals often find slow diffusion of heavy/bulky molecules specifically in the field of fine chemicals and the upgrading of crude oils/vacuum gas oil/residue[6-9]. But severe restriction of diffusion due to the microporous structure of zeolite impedes their use in reactions involving heavy molecules[10]. Most of the industrially used nanoporous materials, Y-zeolite seeks a special attention with 0.74nm diameter has found wide applications in catalysis due to its high acid strength, activity, shape selectivity and stability at high temperature[11-13].

The addition of Nano sized pore network inside the zeolite may help to reduce the diffusion limitation problem. Thus, hierarchal zeolite synthesis seeks attraction among materials as well as catalysis researchers. To date, certain promising techniques have been developed for the hierarchical porous zeolite, mostly were mesoporous zeolites. The approaches used to create the mesoporosity are as follows: (1) Dematalization includes dealumination and desilication[14, 15]which involves post synthetic treatment to create the mesoporosity (2) framework crystallization of synthesized mesoporous aluminosilicates using the steam aided crystallization methods (3) Assembly of nanosized zeolite crystals in which mesopores generated among the crystals.

2 Experimental section

2.1 Synthesis of catalyst

In this study, modified form of the CT-421 obtained after the post treatment with acid, base and organic template hexadecyl trimethy ammonium bromide (CTAB). Modified catalyst name in this study is M-Y.

For post treatment of CT-421, 5g of Na-Y Zeolite (CT-421) is mixed with 50ml of water to make slurry. Few drops of dilute nitric acid used to maintain the PH of the solution at 5.7. After that use 6 miliequivalent of 10% citric acid solution per 1 gram of zeolite and stir it at ambient temperature for 70 min. Vacuum filtration used to separate the solid from the liquid with washing from the demineralized Water. Now dry the separated solid catalyst for 1hr at ambient temperature. After that, made the solution of 0.45g NaOH and 2.6g of CTAB in 50ml water and mix this solution in solid catalyst using stirring. Then pour this solution in PP bottle and put it in oven for 22h at 100ºC. Again filtration starts to separate the solid catalyst from water and place the sample inside oven at 100ºC for overnight drying. Place, prepared zeolite sample inside the furnace for calcination to remove the organic template. Start heating from 25ºC to

120ºC at the rate of 2ºC/min, 120ºC to 550ºC at the rate of 5ºC/min. Stay inside the sample at 550ºC for 5hr. Before proceed to the testing studies, sample needs to be ion exchanged with 0.1 mol/L ammonium nitrate solution.

2.2 Characterization of hierarchical zeolite

Crystallinity of as synthesized samples under study determined by using the X-ray power diffraction. Textural property of the sample catalyst samples measured by using Micromeritics ASAP-2020 analyzer with the help of nitrogen adsorption at 77K. Before physisorption of nitrogen, catalyst sample needs to be degasified at 220ºC for 3 hr. Relative pressure (P/Po) 0.06 to 0.3 used for the calculation of surface area of the catalysts. Silica to alumina (SiO_2/Al_2O_3) ratio determined by using the inductive coupled plasma spectrometer (ICP). Scanning electron microscopy (SEM) used to measure the shape and size of the catalyst. Before the measurement, samples needs to be coated with platinum thin layer. Further, NH_3 TPD used to study the acidity of the modified catalyst.

3 Results and discussions

In this study, XRD have been performed to analyze the structural changes occurred during the acid, base and template treatment. From Figure 1, we can see the similarity of the XRD patterns between the two zeolites but apparent decrease in crystallinity in treated catalyst M-Y also observed. This structural degradation of the modified zeolite due to the severity in the treatment conditions. On the other hand, intracrystalline mesoporosity developed as a result of such treatment as evident by Figure 2. From Figure 2, low angle range peak is detected which is not observed in parent zeolite CT-421. This low angle peak indicates some level of ordered pore arrangement in the mesostructured zeolites. Nitrogen adsorption-desorption data is used for further evidence of the mesoporosity in the treated zeolite.

For treated zeolite M-Y, H4 type of hysteresis loop indicates the presence of both micro and mesopores. But in the parent zeolite CT-421, type 1 isotherm observed which is very different from the treated catalyst. Further, total surface area (internal and external) and pore volume calculated by using the BET method. BET analysis shows the significant increase in the external surface area from 47 to 503 m^2/g in the treated zeolite whereas micropore area decreases from 602 to 274 m^2/g. On the other hand, total volume increases from 0.35 cm^3/g in parent zeolite to 0.59 cm^3/g in treated catalyst. From BET analysis, it is evident that mesoporosity is created inside the treated zeolite.

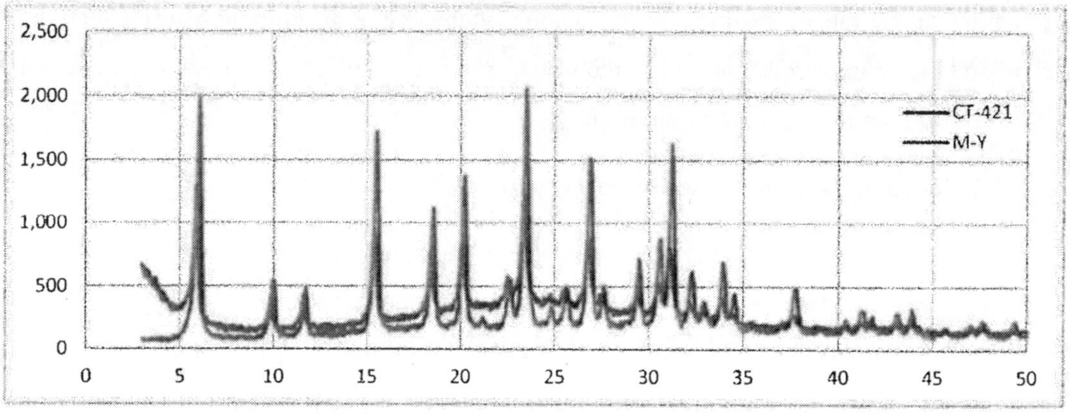

Figure 1. Wide angle XRD pattern of parent CT-421 and modified M-Y zeolite.

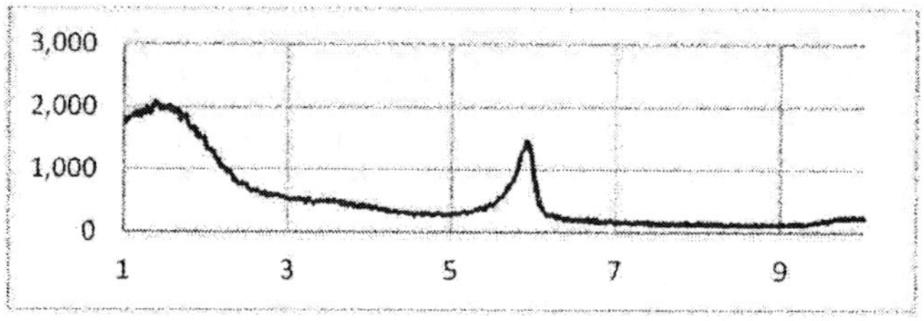

Figure 2. Low angle XRD pattern of treated M-Y zeolite.

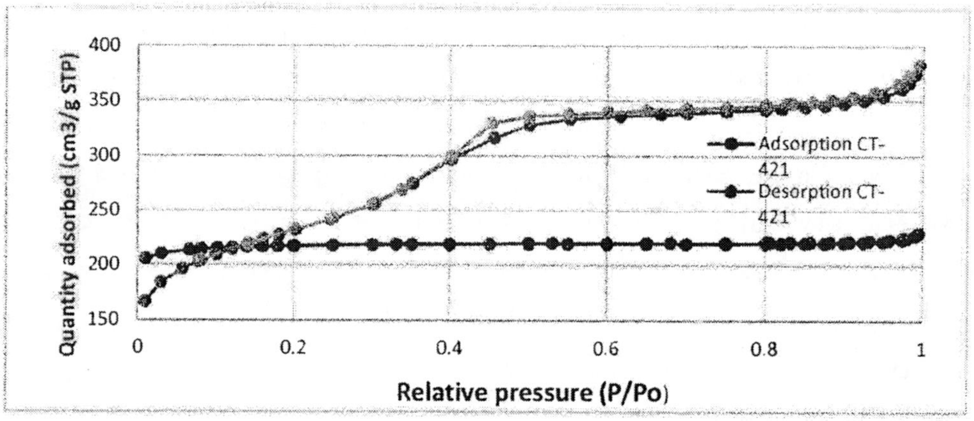

Figure 3. N_2 adsorption-desorption isotherm of parent zeolite CT-421 and treated zeolite M-Y.

4 Conclusion

In reality, FCC catalytic process is 75 years old but research in this area is still very much alive. Recent feedstock changes in the demand for middle distillates, gasoline and propylene need further improvement in the process side as well as catalyic. As we know that FCC is one of the largest process is used in the world, any improvement in the efficiency of the process (i-e coke making) will multiply to large factors to save millions of ton CO_2 per year. Hierarchical zeolites having strong acidity and uniform mesoporosity have been synthesized by post synthetic treatment of zeolite in the presence of acid, base and surfactant. This methodology already applied successfully for synthesis of well-defined mesoporous materials. However, mechanism of intermediate structure and mesostructuring process still needs to be understand. A detailed comprehension about the mesostructuring process can lead to target post synthetic process in a right way which yield better material having direct influence on catalytic applications, especially in the conversion of bulky molecules.

5 References

1. Gary, J.H., G.E. Handwerk, and M.J. Kaiser, *Petroleum refining: technology and economics.* 2007: CRC press.
2. Ahmed, M.H., et al., *Stability Assessment of Regenerated Hierarchical ZSM-48 Zeolite Designed by Post-Synthesis Treatment for Catalytic Cracking of Light Naphtha.* Energy & fuels, 2017. **31**(12): p. 14097-14103.
3. Galadima, A. and O. Muraza, *Stability improvement of zeolite catalysts under hydrothermal conditions for their potential applications in biomass valorization and crude oil upgrading.* Microporous and Mesoporous Materials, 2017. **249**: p. 42-54.
4. Jamil, A.K., et al., *Hydrothermally stable acid-modified ZSM-22 zeolite for selective propylene production via steam-assisted catalytic cracking of n-hexane.* Microporous and Mesoporous Materials, 2018. **260**: p. 30-39.
5. Armor, J.N., *New catalytic technology commercialized in the USA during the 1980's.* Applied catalysis, 1991. **78**(2): p. 141-173.
6. Čejka, J. and S. Mintova, *Perspectives of micro/mesoporous composites in catalysis.* Catalysis Reviews, 2007. **49**(4): p. 457-509.
7. Choi, M., et al., *Stable single-unit-cell nanosheets of zeolite MFI as active and long-lived catalysts.* Nature, 2009. **461**(7261): p. 246.
8. Meng, X., F. Nawaz, and F.-S. Xiao, *Templating route for synthesizing mesoporous zeolites with improved catalytic properties.* Nano Today, 2009. **4**(4): p. 292-301.
9. Pérez-Ramírez, J., et al., *Hierarchical zeolites: enhanced utilisation of microporous crystals in catalysis by advances in materials design.* Chemical Society Reviews, 2008. **37**(11): p. 2530-2542.
10. Christensen, C.H., et al., *Catalytic benzene alkylation over mesoporous zeolite single crystals: improving activity and selectivity with a new family of porous materials.* Journal of the american chemical society, 2003. **125**(44): p. 13370-13371.
11. Liu, B., et al., *Template synthesis of the hierarchically structured MFI zeolite with nanosheet frameworks and tailored structure.* New Journal of Chemistry, 2014. **38**(9): p. 4380-4387.
12. Soler-Illia, G.J.d.A., et al., *Chemical strategies to design textured materials: from microporous and mesoporous oxides to nanonetworks and hierarchical structures.* Chemical reviews, 2002. **102**(11): p. 4093-4138.
13. Srivastava, R., M. Choi, and R. Ryoo, *Mesoporous materials with zeolite framework: remarkable effect of the hierarchical structure for retardation of catalyst deactivation.* Chemical Communications, 2006(43): p. 4489-4491.
14. Mei, C., et al., *Regular HZSM-5 microboxes prepared via a mild alkaline treatment.* Journal of Materials Chemistry, 2008. 18(29): p. 3496-3500.
15. Pérez-Ramírez, J., et al., *Toward Functional Clathrasils: Size-and Composition-Controlled Octadecasil Nanocrystals by Desilication.* Angewandte Chemie, 2008. **120**(41): p. 8031-8035.

Crude oil–to–chemicals in the global petrochemical industry; Applications of Nanotechnology

Tawfik A. Saleh

Chemistry department, KFUPM

tawfik@kfupm.edu.sa

Abstract

Several crude oil–to–chemical (COTC) technologies being constructed or planned worldwide have the potential to reshape the global petrochemical industry in years to come. COTC configures a refinery to produce maximum chemicals instead of traditional transportation fuels.

There are several processes among them some processes used catalysts. Catalysts made from nanoparticles have a greater surface area to interact with the reacting chemicals than catalysts made from larger particles. The larger surface area allows more chemicals to interact with the catalyst simultaneously, which makes the catalyst more effective. This increased effectiveness can make a process such as the production of diesel fuel from coal more economical, and enable the production of fuel from currently unusable raw materials such as low grade crude oil. This presentation will highlights the recent trends in such COTC technologies with highlights to the impact of nanotechnology.

Keywords: Oil, Petrochemicals; Catalysts; Nanotechnology

29th Annual Saudi-Japan Symposium – 2019

Technology in Petroleum Refining & Petrochemicals

Venue: KFUPM Dhahran Techno Valley, Innovation Cluster Bldg., Auditorium 1111: Nov. 3-4, 2019

Day One: Sunday, November 3, 2019

8:30 OPENING REMARKS SESSION ***Chairman: Dr. Hassan Al-Asiri***

1. *H.E. Dr. Sahel Abuduljauwad, Rector, KFUPM*
2. *Mr. Eiji Hiraoka, Senior Executive Director, JCCP*
3. *Mr. Ahmed Al-Khowaiter, CTO, Saudi Aramco*
4. *Dr. Keiichi Tomishige, Professor, Tohoku University*

SESSION ONE FUTURE REFINING TRENDS ***Chairman: Dr. Keiichi Tomishige***

9:00 1. **Keynote: Technical trends and research topics in the Japanese refining industry**
Dr. Jinichi Igarashi, JX Nippon Research Institute, Ltd.

9:30 2. **An economic perspective of light naphtha upgrading**
Dr. Yaming Jin, Dr. Omer R. Koseoglu, Saudi Aramco

9:50 *Coffee Break*

SESSION TWO NEW CHALLENGES: CHEMICALS/FUELS ***Chairman: Dr. Omer R. Koseoglu***

10:15 3. **Keynote: Recent advances in zeolite catalysis for production of petrochemicals**
Dr. Suheil Abdo, Honeywell UOP, USA

10:45 4. **Catalyst technology to maximize light olefin yield in FCC unit**
Dr. Takaki Mizuno, Petroleum Refining Catalysts Research Center, R&DC, JGC C&C Lt

11:05 5. **Recent advances and overviews of Ni(Co)Mo supported mesoporous silica catalysts for HDS activity of liquid fuel,** *Dr. Saheed Ganiyu, Dr. K. Alhooshani, Dr. A. Tanimu, KFUPM/CHEM*

11:25 6. **Effects of titania coatings on HDS catalysts: insights from first principles calculations**
Dr. Hiroki Iriguchi, Central Technical Research Lab., JXTG Nippon Oil & Energy Corp.

11:45 *Prayer & Lunch Break*

SESSION THREE NEW CHALLENGES: PROCESS/FUELS ***Chairman: Dr. Mamdouh Al-Harthi***

13:00 7. **Application of inferential models to crude distillation process: integrating laboratory feedback in closed-loop quality control,** *Dr. Rashid Ansari, Saudi Aramco*

13:20 8. **Catalytic dealkylation of alkyl polycyclic aromatic hydrocarbons towards innovation of petroleum refinery,** *Dr. Naonobu Katada, Center for Research on Green Sustainable Chemistry, Tottori University*

13:40 9. **Selective aerobic oxidation of hydrocarbons by N-hydroxyphthalimide-based catalyst for cetane enhancement,** *Dr. Maryam Altaher, C. Kalamaras, M. Alqahtani, Saudi Aramco*

POSTER SESSION

14:00 ***14 Poster contributions from KFUPM Chemistry, CENT, CRP and Jazan University-CHE***
- Reviewing and commending excellent poster presentation (5-min each)

15:30 *Day One Ends*

Each presentation includes 5-minutes Q&A http://catsymp.kfupm.edu.sa

29th Annual Saudi-Japan Symposium – 2019

أرامكو السعودية
saudi aramco

Technology in Petroleum Refining & Petrochemicals

Venue: KFUPM Dhahran Techno Valley, Innovation Cluster Bldg., Auditorium 1111: Nov. 3-4, 2019

Day Two: Monday, November 4, 2019

SESSION FOUR NEW CHALLENGES: CATALYST TECHNOLOGY -1 *Chairman: Dr. Khalid Alhooshani*

8:00		*Registration & Coffee Break*
8:20	**10.**	**Catalytic technology for production of C4 chemicals via erythritol from biomass** *Dr. Keiichi Tomishige, Dept. Applied Chemistry, Tohoku University*
8:40	**11.**	**Effects of residual catalyst structure and electronic environment on polyethylene crystallization and degradation,** *Dr. Muhammad Atiqullah, KFUPM/CRP*
9:00	**12.**	**Sintering– and coking– resistant core-shell nanoparticles in dry reforming of methane** *Dr. Noor Al-Mana, Saudi Aramco*
9:20	**13.**	**Oxidative dehydrogenation of propane to propylene over** $VOx/CaO\text{-}\gamma Al_2O_3$ *Dr. M. Mozahar Hossain, KFUPM/CHE, Saudi Aramco*
9:40		***Coffee Break***

SESSION FIVE NEW CHALLENGES: CATALYST TECHNOLOGY -2 *Chairman: Dr. Ali Alzaid*

10:00	**14.**	**Effect of catalyst dilution and binders on the alkylation activity of USY zeolites** *Dr. Mohammed R. Aldossary, Vladimir Zholobenko, Arthur Garforth, Saudi Aramco*
10:20	**15.**	**Heavy naphtha reforming of** C_{12} **over variable-morphological MFI zeolites,** *Dr. Oki Muraza, M. Sanhoob, E. Shafei, M. Al-Bahhar, A. Al-Jishi, K. Choi, A. Khan, KFUPM/CHE-CENT, Saudi Aramco*
10:40	**16.**	**New generation of naphtha reforming catalysts,** *Dr. Ali Jahel, Axens, France*
11:00	**17.**	**The effect of zeolite crystal structure on hydrocarbon conversion,** *Dr. Mohammad Al Bahhar, E. Al-Shafei, A. Al-Jishi, O. Muraza, M. Sanhoob, A. Jamil, K-H. Choi, A. Alnaser, H. Al-Badairy, Saudi Aramco*
11:20	**18.**	**Machine learning approach for prediction of reaction yield with simulated catalyst parameters,** *Dr. Akira Yada, National Institute Adv. Industrial Science & Technology (AIST)*
11:40		*Prayer & Lunch Break*

SESSION SIX CRUDE TO CHEMICALS/CRACKING *Chairman: Dr. Naonobu Katada*

13:00	**19.**	**Keynote: Advancing the frontier in shape selective catalysis to enable the production of petrochemicals in the modern refinery,** *Dr. Eelco Vogt, Advisor Albemarle, Utrecht University*
13:30	**20.**	**Upgrading oil using a non-conventional charged droplets approach** *Dr. Chanbasha Basheer, KFUPM/CHEM*
13:50	**21.**	**Catalytic cracking of pentene to propene and ethene over ZSM-5 zeolite** *Dr. Abeer Ariah, Dr. Arthur Garforth, Saudi Aramco*
14:10	**22.**	**Application of hierarchical zeolites in FCC catalyst,** *Dr. Shakeel Ahmed, KFUPM/CRP*
14:30	**23.**	**Crude oil–to–chemicals in the global petrochemical industry: Applications of nano-technology,** *r. a 'fi ________, KFUPM/CHEM*
14:50		***Closing Remarks, Symposium Ends*** D. T w k A Saleh

Each presentation includes 5-minutes Q&A *Program as of October 13, 2019*

For Registration (Free): http://catsymp.kfupm.edu.sa/register.htm **Phone 013 860 2029**

29th Annual Saudi-Japan Symposium – 2019

ارامكو السعودية
Saudi Aramco

Technology in Petroleum Refining & Petrochemicals

Venue: KFUPM Dhahran Techno Valley, Innovation Cluster Bldg., Auditorium 1111: Nov. 3-4, 2019

Day One: Sunday, November 3, 2019

POSTER SESSION (TIME: 14:00-15:30)

01. **Synthesis of Nickel/zeolite catalyst by chemical deposition: pathway to highly dispersed and stable Nickel catalyst**
Yahia Hakmi, KFUPM-Chemistry.

02. **Facile synthesis of ultrathin interconnected carbon nanosheets as a robust support for small and uniformly-dispersed iron phosphide for the hydrogen evolution reaction**
Munzir Suliman, KFUPM-CENT/Chemistry.

03. **Investigating the effect of Ammonium Fluoride (NH4F) Media in Synthesis and Characterization of a Metal Containing Zeolite Catalyst Material**
Alhassan Ibrahim, KFUPM-Chemistry.

04. **Sodium Octanoate Functionalized Graphene Oxide (SO-GO): A Green Corrosion Inhibitor for Oil Well Acidizing Environment**
Kabiru Haruna, *KFUPM-Chemistry.*

05. **Pd-NHC - Catalyzed Effective Synthesis of Biaryl ketones and Biaryl diketones by Carbonylative Suzuki-Miyaura Coupling Reactions**
Waseem Mansour, *KFUPM-Chemistry.*

06. **Modified ZSM-5 Supported NiMo-S for Hydrodesulfurization application**
Umar Mustapha, KFUPM-Chemistry.

07. **Novel mesoporous cobalt iron sulfide for efficient hydrogen evolution reaction**
Dr. Alaadin Adam, KFUPM-CENT.

08. **Effect of Si/Al Ratio of MFI-type Zeolite Catalyst on Light Naphtha Aromatization**
Mohammed Ellouh, KFUPM-CRP.

09. **Synthesis of Methyl t-Butyl Ether using low cost activated carbon-based catalyst**
Dr. M. Ashraf Ali, ChE-Jazan University.

10. **Carbon Nanofiber-Doped Zeolite as Support for Molybdenum Based Catalysts for Enhanced Hydrodesulfurization of Dibenzothiophene**
Islam Ali, T. A. Saleh, KFUPM-Chemistry.

11. **Hydrodesulfurization of Dibenzothiophene over NiMo Dispersed on Carbon-TiO2 Composites**
Umar Cheche Abubakar, KFUPM-Chemistry.

12. **SOMC-generated single-site VOx species on silica for ODH of propane**
Dr. Samir Barman, KFUPM-CRP.

13. **Effect of Screw Speed and Material Type on the Energy Consumption of Multiscrew Extruders**
Dr. Mansour Albareeki, KFUPM-CRP.

14. **$LaCO_3OH$ Nanoprisms: Photoluminescence and toxic NOx reduction properties**
Dr. Hassan Zahir, KFUPM/CoRE-RE